THE ENCYCLOPEDIA OF MOON MYSTERIES

Secrets, Conspiracy Theories, Anomalies, Extraterrestrials and More

Victoria Constance Briggs

The Encyclopedia of Moon Mysteries
By Victoria Constance Briggs

Adventures Unlimited Press

ISBN: 978-1-948803-10-6

Published by:
Adventures Unlimited Press
One Adventure Place, Box 74
Kempton, Illinois 60946 USA
auphq@frontiernet.net

www.adventuresunlimitedpress.com

Cover illustration by Mister Sam Shearon
www.mistersamshearon.bigcartel.com

Acknowledgements

Thank you…

To my husband Ghobad Heidari, for your endless love and tireless support; without whom, this work would not have come to fruition.
To my son Kion Heidari, for your encouragement in my pursuit of the metaphysical and the esoteric.
To my daughter Shireen Heidari, for your support and for convincing me to pursue my dreams.
To David Hatcher Childress and Adventures Unlimited Press, who believed in this book.
To Eva Brucia for your wonderfully creative illustrations.
To Pearl Jimelle Beavers for your imaginative artwork.
To my brother Marcus Briggs, for always being there for me, and for listening.
To Barbara Murphy for your friendship and professional advice.
To Lola Tam for your professional opinions and encouragement.
To Mamatha Rosenbaum for your support and friendship.
To Don Wilson for your inspiration…wherever you are!
To the astronomers of old that pursued knowledge.
To the astronauts who had the nerve to explore…our mysterious Moon.
…to you all…I am grateful!

I would also like to acknowledge, the Library of Massachusetts Historical Society, the National Aeronautics and Space Administration, and the San Diego Public Library (Carmel Valley Branch).

THE ENCYCLOPEDIA OF MOON MYSTERIES

Secrets, Conspiracy Theories, Anomalies, Extraterrestrials and More

To my Mother,
Arnita Elizabeth Holly Briggs
With Love
Thank you for Everything!

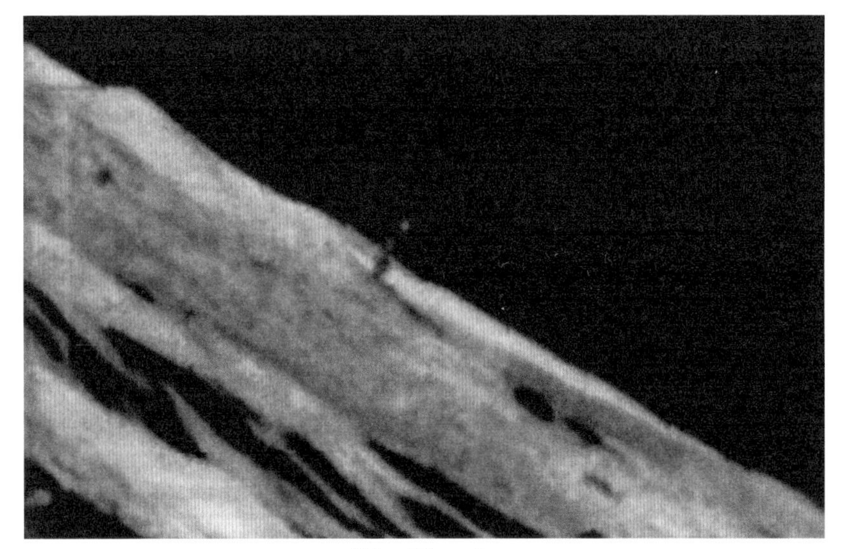

The Shard.

THE ENCYCLOPEDIA OF MOON MYSTERIES

Secrets, Conspiracy Theories, Anomalies, Extraterrestrials and More

INTRODUCTION

Our moon is an enigma. The ancients viewed it as a luminous orb, there to brighten the night sky and act as a beacon in the darkness. Some cultures even worshipped it as a god. When the telescope was invented, people took to studying the moon. Soon, maps of the Moon were created, and the speculation arose that the Moon was similar to the Earth, with mountains, valleys, seas; and the belief that it was inhabited. Strange lights and shadows were seen moving across the Moon, supporting their theory that the Moon is a vibrant, populated place. As the years passed by, and telescopes became more powerful, scientists began to realize that the Moon could not sustain life. Yet, the strange lights and shadows on the Moon persisted. Just who or what was causing the commotion is still a mystery.

This book is a compendium of odd events and strange and unusual activities on and surrounding the Moon. Inside, you will find information that suggests that the Moon is home to extraterrestrials, theories that it is not a natural satellite, sightings of anomalous lights, and claims that NASA astronauts saw extraterrestrial spaceships and unnatural constructions on their missions. There is a story of an Apollo mission being saved by extraterrestrials, rumors of clandestine bases, a legend stating that a part of the kingdom of Atlantis was once located on the Moon, and theories of cities beneath the lunar surface. There is an account of a government employee using remote viewing as a means of seeing what is on the Moon and more. You will also find information on famous astronomers, scientists and other notables and their discoveries about the Moon, along with interesting facts.

The mysteries surrounding our Moon may be just a sample of what we may find as we continue our travel into the cosmos. We have so much to learn about the universe in which we have found ourselves. Perhaps one day a door will open and all questions will be answered. Until then, it is a wonderful and fascinating journey into the unknown. Enjoy the ride!

"There is more to our universe than we have been taught. We need to look deeper and search further if we want to know the truth of who we are, where we come from, and what lies beyond."

—*Constance*

The Russian Zond 3 photograph showing what seems to be a huge tower.

"The most beautiful thing we can experience is the mysterious. It is the source of all true art and science."
–Albert Einstein

A

(2001) A Space Odyssey

A movie by director, producer and screenwriter **Stanley Kubrick**, starring Keir Dullea and Gary Lockwood. The screenplay was cowritten by Kubrick and Arthur C. Clarke. The story involves the locating of an ancient relic left on the Moon by advanced **extraterrestrials** millions of years before in order to track mankind's development and to signal them when human's had progressed to space travel. The movie reflects the theory that there may be advanced extraterrestrials monitoring mankind's progress into space from the Moon.

Abduction(s)

The act of being taken against one's will by **extraterrestrials** to an undisclosed location. There are claims from people around the world of having been abducted by extraterrestrials, more commonly referred to as "alien abductions." People that have been abducted are referred to as "abductees." They are also referred to as "experiencers." Abductees often experience physical examinations at the hands of their captors. There are additional reports of the examinations being both mental and physical. Reportedly they often involve the reproductive system. In some cases, implants are placed inside of the abductees, presumably to later track them on earth. As a result, some abductees claim to have been abducted multiple times. The extraterrestrials most often identified in these scenarios are known as the "**Zetas**" or the "**Greys**." There have been reports of some abduction experiencers allegedly having been taken to the Moon. Some recalled that while on the Moon, they were taken to underground bases, where physical examinations ensued. See **Extraterrestrial Bases**.

Adam

The first human created by God in the creation beliefs of Judaism, Christianity and Islam. In the Bible book of Genesis 2:7 (*New International Version*), it states, "Then the Lord God formed a man from the dust of the ground and breathed into his nostrils the breath of life and the man became a living being." Adam is also "Adapa," the first human created by the gods of ancient Sumerian beliefs. These gods are known as the **Anunnaki**. The Sabaean culture maintained that Adam originated on the Moon and was brought to Earth by his creators the Anunnaki. Adam, per some ancient tales, revered the Moon and taught others to as well. These ancient beliefs about Adam's association with the Moon have led to the idea that the Moon is really the planet **Nibiru.** Another theory is that the Moon is a disguised spaceship from Nibiru. Nibiru is home to the Anunnaki. For this reason, some believe that Adam may have considered the Moon home and the home of the gods. In expressing the beliefs of the Sabaeans, Moses Maimonides (influential medieval Jewish philosopher) wrote, "They deem Adam to have been an individual born of male- and female-like any other human individuals, but they glorify him and say that he was a prophet, the envoy of the moon."

Per the Sumerian tale, Adam was created by the two gods **Enki and Ninki** (female) who were a part of the Anunnaki. They are believed to have created a slave race to mine Earth for

gold. In the article titled "The Scariest Book of All Time: Cosmological Ice Ages a Short Explanation," Henry Kroll, the author of the book *Cosmological Ice Ages,* writes, "The fact that Adam may have been preaching a worship of the Moon is just more evidence of his possible birthplace and the location of his acknowledged creators."

Adamski, George
1891–1965

A Polish born American citizen that later became one of the world's most famous extraterrestrial **contactees**. Adamski served in the army before and during World War 1. He was a philosopher, author, **extraterrestrial** researcher-investigator, photographer and astronomer. During the 1950s Adamski became known for his interest in **spaceships** (the term **UFO** had not become popular during this period) and spent a great deal of time searching for them. He started his research with only a six-inch telescope. He documented many of his sightings of spaceships and substantiated his findings through photographs and video footage. Adamski described several of the crafts he saw as being either round or **cigar-shaped**. He also had witnesses to vouch for him, which lent credibility to his sightings. In his book *Flying Saucers Have Landed*, Adamski tells his story.

Adamski's life changed considerably when on one fateful day, as he tells it, he came face to face with a humanoid **extraterrestrial** that communicated with him. This being was said to be a male, of Nordic (humanoid extraterrestrials from the Pleiades) origins and from the planet Venus. He had come to Earth to bring a message to mankind in hopes of convincing them not to use the atomic bomb. He instructed Adamski to relay this information to the people of Earth in order to keep peace in the world and to limit the negative domino effects that the setting off of an atomic bomb would have in the galaxy. Thus, Adamski became the human voice of the **extraterrestrial** visitors.

One area of concentration for Adamski was the Moon. According to Adamski, during some of his observations, he witnessed spaceships flying in the direction of the Moon. In an article published in *Fate Magazine* in 1951, Adamski spoke of seeing spaceships that appeared to land on the **nearside** of the Moon. He noticed that others flew towards the **far side** and would disappear. In the article Adamski stated, "I figure it is logical to believe that spaceships might be using our moon for a base in their interplanetary travels."

Adamski also claimed that he had traveled with his extraterrestrial companions on board a spaceship on more than one occasion. During two of those trips they circled the Moon. Adamski was given information about the Moon from the extraterrestrials while they were in orbit. In his writings, Adamski shared what he was told and what he had observed. According to Adamski, the extraterrestrials relayed to him that the Moon has air. They conveyed to him that the nearside of the Moon does not reach the high temperatures that researchers on Earth imagine. He also claimed that there is a localized area on the Moon where life exists, including he says, a form of **vegetation**, animals and humans. Additionally, Adamski shared what he had witnessed when observing the Moon through telescopic equipment on the spacecraft. He stated that some the craters were actually canyons that are encircled by mountain ranges. He also claimed to have seen evidence that showed that at some point on the Moon, **water** was once there. The extraterrestrials informed Adamski that water could be found on the far side of the Moon and also beneath some of the mountains on the nearside of the Moon. Adamski described some areas of the lunar **surface** as being made up of a light, dusty type of material, while other areas were more of a granular substance. On August 23, 1954, Adamski for the second time took a trip to the Moon. There he claims to have observed enormous air docks on crater floors designed to hold large spacecraft. Adamski was informed that visitors to the Moon had to first depressurize, in order for their bodies to become adjusted to the lunar environment. On the far side, Adamski says that he witnessed mountains covered with snow. He described seeing several communities on the sides of hills and mountains and in some low lying areas. He also observed a large town. He described seeing hangars shaped like domes for spaceships coming in that were carrying supplies. There was a barter system in place where supplies were traded for minerals from the Moon. Adamski spent the remainder of his life seeking to educate mankind about life in outer space. His goal was to help people understand that humans are not alone, but are a part of a universal community, which included extraterrestrials on the Moon.

Adapa

The first created human in ancient Sumerian beliefs. According to lore, he was created by the **Anunnaki**, who some believed were gods that operated from the Moon. See **Adam**.

Aega

A moon goddess of ancient Greek mythology. Her mother was Gaia, the goddess of the Earth. Known for her beauty, Aega radiates the mysterious light of the Moon.

African Moon Legend
See **Zulu**.

Age of the Moon

The age of the Moon according to most modern-day scientists is somewhere around 4.6 billion years. This means that the Moon is older than the Earth which is estimated to be 4.5 billion. This discrepancy in age has led some researchers to consider the possibility that the Moon was originally created outside of our

universe and in some mysterious way, came to be in Earth's orbit.

Aine

Celtic fairy moon goddess, whose name means "radiant." Aine lights up the darkness of the night. She and her two sisters ride their horses when the Moon is high and play in the waters of the divine.

Alaje of the Pleiades

An individual that claims to be an incarnated **extraterrestrial** from the Pleiades. He communicates via the Internet and has sustained a large audience. He is considered by many to be a genuine extraterrestrial person. Alaje is said to have incarnated on Earth to spread enlightenment, teach love and tell humanity the truth about the Earth's history. He teaches that love is powerful and is the answer to problems affecting the Earth's residents.

Alaje states that there are cities on the Moon that are located in both the interior and exterior of the Moon's **far side**. He holds that people have been living there since the 1940s. He also maintains that the Moon was brought into

Earth's orbit ages ago, during a warring period on Earth. He states that during that time, the Moon was used as a **base**. According to Alaje, the Moon is artificial and that is the reason that we only see the **nearside**; it is designed with a technology that prevents it from turning, so that the **structures** and spacecraft remain hidden from the people of the Earth.

Aldrin, Buzz
(Edwin Eugene Aldrin, Jr.)
1930—

A former NASA **astronaut**, Aldrin was the Lunar Module Pilot on the historic ***Apollo 11*** mission. He was the second man to set foot on the Moon. Aldrin once served as an Air Force colonel, is a pilot and holds a doctorate of Science in Astronautics. He also served onboard *Gemini 12* where he set a record for extravehicular activity (EVA). On July 20, 1969, as he stepped onto the Moon, he famously described the scene around him saying, "Beautiful, beautiful, magnificent desolation." There are a number of mysterious stories surrounding Aldrin and the *Apollo 11* mission. There are rumors of Aldrin and **Neil Armstrong** witnessing **extraterrestrial** spaceships parked on the Moon. They allegedly transmitted on a secret line that they were being watched by whoever was in the ships. There was also an incident with a **UFO** on the way to the Moon, which Aldrin recounts in his first autobiography, *Return to Earth*. It is believed by some that what was seen that day was a piece of the *Saturn V* rocket.

Alien Abductions
See **Abductions**.

Alphonsus (Crater)

An impact **crater** that dates back to the pre-Nectarian period. It is seventy-three miles in diameter and is located on the eastern edge of Mare Nubium. There, astronomers have reported seeing mysterious, unexplained lights. On November 3, 1958, Russian astronomer **Nikolai Kozyrev** of the Crimean Astrophysical Observatory photographed what appeared to be red lighting near Alphonsus. According to Kozyrev the light appeared to move about and then vanished after an hour. It has been speculated that what Kozyrev witnessed was the result of gas spewing forth from the crater. However, researchers speculate that it was not related to natural activity, but something far more mysterious. Many of the reported sightings inside the crater involve red luminosities, mysterious streaks of light and glowing lights. Reportedly, these strange illuminations appear and disappear in various parts the crater. A list of **transient lunar phenomena** inside the Alphonsus crater follows.

1958, November 19, a strange cloud-like formation was observed hovering over the crater's central mountain area for thirty minutes.

1958, November 19, an area in the crater's central peak was illuminated in red.

1958, November 22, an odd, unexplained grayish patch was seen inside the crater.

1958, December 19, an area the crater's central peak glowed red.

1958, February 18, a small red light was seen inside the crater.

1959, October 23, a red glowing spot was observed within the crater.

1960, January 6, a red point of light was seen.

1964, October 27, A ruddy area was perceived at the base the crater's central peak during a period when sunlight was beaming down upon it.

1966, May 27, dim red areas of light were seen and monitored for fifty minutes.

1966, September 22, there were a series of faint glowing lights inside the crater that lasted for an hour.

1967, February 17, a blinking light was seen just inside of the southwest floor the crater. It was monitored for twenty-five minutes.

1967, February 19, for ten minutes, a red light was observed in the same area as the blinking light that was seen on February 17.

Alpine Valley
See **Vallis Alpes**.

Anahita

A Persian goddess of the Moon. Her name stands for "uncorrupted." She symbolizes the purification of the Moon and the cosmos.

Anaxagoras of Clazomenae
c. 500 – c. 428 BC

A renowned Greek philosopher whose main area of study was cosmology. Anaxagoras was fascinated with understanding the universe and how it works. As a teacher in Athens, Anaxagoras became famous for his concepts about the cosmos and the Milky Way. He was popular with many scholars from that time, including Greek tragedian Euripides as well as the Greek statesman Pericles. During a period when man was trying to understand his place in the universe, Anaxagoras provided them with answers and ignited a hunger in them that made them eager to learn more. His knowledge of the cosmos can be seen in how we perceive the universe today.

It was Anaxagoras that first explained some of the mysteries surrounding the Moon, including why the Moon shines at night. In his work *A Refutation of All Heresies*, he writes, "The moon does not shine with its own light, but receives its light from the sun." He also explained the reasons for both the Sun and the Moon's eclipses. Again in, *A Refutation of All Heresies* he states, "Eclipses of the sun take place at new moon, when the moon cuts off the light." At a time when many people still worshiped various objects in the cosmos, Anaxagoras was put on trial for irreverence when it came to his theories about the Sun's and Moon's function in the universe. He was eventually forced into retirement and left Athens.

Anaxagoras believed that the Moon's **surface** is made up of mountainous terrain, plains and valleys much like Earth and that there were people living on the Moon. He also made comments about there being a time in Earth's history when there was no Moon in the sky. Anaxagoras' statements about a Moonless Earth have been used in arguments in more recent times among ufologists who claim that the Moon was brought into our solar system from elsewhere in the universe. See **Pre-Lunar Earth.**

Ancient Moon Map
See **Lunar Map at Knowth**.

Anders, William A.
1933–

A former NASA **astronaut**, Anders served aboard ***Apollo 8*** as the Lunar Module Pilot. In reminiscing about his trip to the Moon, Anders once commented, "The thing that impressed me the most was the lunar sunrises and sunsets. These in particular bring out the stark nature of the terrain...The horizon here is very, very stark, the sky is pitch black and the Moon is quite light and the contrast between the sky and the Moon is a vivid dark line." On December 24, 1968 while working under the eerie, darkness of the Moon's sky, Ander's snapped a picture of Earth that is now known as the iconic, "***Earthrise***" photograph. It has been hailed as "the most influential environmental photograph ever taken," by acclaimed wilderness photographer Galen Rowell. It remains the greatest picture ever taken from the **Apollo** moon missions.

Angels

The custodians of the cosmos. Angels can be found in beliefs in cultures and religions around the world. There are angels of the Sun, angels of the stars and angels of the Moon.

Six Russian **cosmonauts** claimed to have seen angels while on missions in space. Reportedly, there were two separate sightings with two separate crews. The story is said to have been smuggled out of Russia in 1985 by a scientist who was defecting. According to the report, the cosmonauts of the *Salyut 7* space station, which included, Vladimir Solevev, Oleg Atkov and Leonid Kizim, were surprised when a blinding, bright orange light suddenly appeared, beaming into the space station. When they looked out of the portholes to locate the light's source, they saw seven angels. The angels were described as "seven giant figures," in human form, complete with halos and wings that were the size of jetliners. Each angel was indistinguishable from the other. The angels are said to have followed the space capsule for ten minutes and then, mysteriously disappeared. Twelve days later the angels appeared again. This time three different cosmonauts saw them. Again, the brilliant light appeared and the three new cosmonauts, by the names of Svetlana Savitskaya, Igor Volk and Vladimir Dzhanibekov saw the gigantic angels. Svetlana Savitskaya was quoted as saying, "We were truly overwhelmed. There was a great orange light and through it we could see the figures of seven angels. They were smiling as though they shared a glorious secret." See **Angels of the Moon**.

Angels of the Moon

In Judaism and Christianity the angel of the Moon is Gabriel. In Islam it is Jibril (who is also Gabriel). Opanni'el YWH is the angel of the orb of the Moon. In the occult, the angel of the Moon is Tsaphiel. The angel Abuzohar is also an angel of the Moon. In Persian mythology the angel of the Moon was Mah. In ancient African beliefs, Yemaya governed the Moon. Asder'el brought the knowledge of the phases of the Moon to the Earth. The Canaanite priestesses were taught the power of the Moon and how to utilize its phases from the angel Sariel.

Animal Experiments

Tests using animals to learn the effects of space on the body after leaving Earth's atmosphere. Several of these experiments were performed in preparation for future manned missions to the Moon. Scientists were seeking to learn the effects in areas such as weightlessness and radiation from the Sun. The United States and the **Soviet Union** opted to perform the tests on animals as opposed to putting human lives at risk. Fruit flies were the first living creatures to be sent into space in 1947. The first monkey was a rhesus monkey named Albert 1, which was launched on July 11, 1948. Both were sent up from the United States on V2 rockets on non-orbital missions. The first animal to be sent into orbit was a mixed breed dog named Laika. Laika was launched via the Soviet *Sputnik 2* mission on November 3, 1957. Laika died while the mission was still in progress. On July 2, 1959, the Soviets sent two dogs and the first rabbit into space. It was the Soviet *Sputnik five* that was

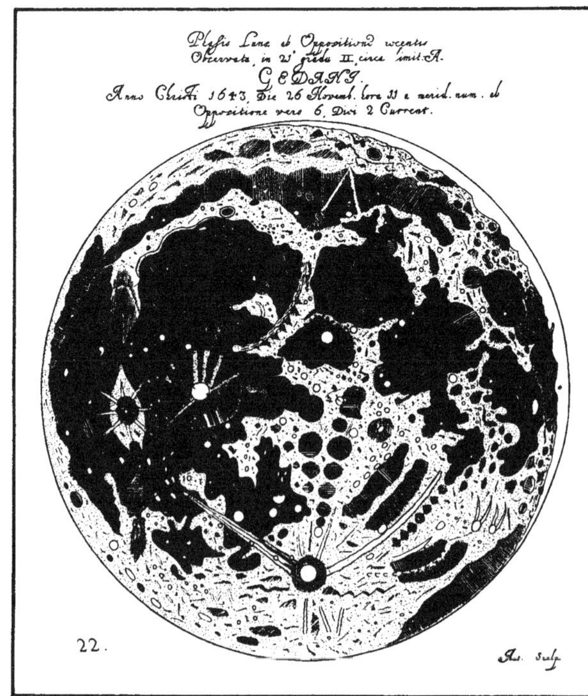

launched on August 19, 1960 that first returned with animals that were still alive from orbit. The animals on that mission included two dogs, a rabbit, 42 mice, two rodents and fruit flies. In addition, a number of other types of animals were launched into space in an effort to perfect space travel for humans. These included rodents, frogs, dogs, cats, monkeys and more. Several of the animals died in space during the tests. Once an animal perishes in space, it remains there forever. It is said that the **astronauts** would see them during their missions, still floating in space.

Anomalies (Moon)

Anomalies have reportedly been seen on the Moon by astronomers dating back to the early 1600s when the telescope was first invented. Later, once mankind progressed to travel into space and then to the Moon, more reports became available of strange sightings of lights on and around the Moon. The anomalies included weird formations and moving shadows, geometric-shaped lights, objects moving across the lunar **surface** and more. Author David Hatcher Childress comments on the topic in his book *Extraterrestrial Archaeology*, page 16, where he states, "Lights, clouds, apparent structures and mysterious streaks have fueled speculation on the mysteries of the Moon for hundreds of years." See **Appendix 2: Anomalous Lights; Craters; Chronological Catalogue of Reported Lunar Events; NASA Technical Report TR-277.**

Anomalous Lights

See **Appendix 2: Anomalous Lights; Transient Lunar Phenomenon.**

Antenna

A researcher studying images of the Moon claimed to have located an antenna protruding out of the lunar **surface**. It was his belief that the antenna may have been connected to an underground establishment placed there by **Moon dwellers**. Skeptics maintain that it is a trick of the light.

Anunit

A Babylonian goddess of the Moon and the evening star.

Anunnaki

(Those of Royal Blood)

Deities of ancient Sumerian beliefs. It has been hypothesized by some lunar researchers that the Moon is a spaceship brought into Earth's orbit by the Anunnaki. According to ancient Sumerian lore, the Anunnaki created Adapa (**Adam**) the first human. This is thought to have possibly occurred on the Moon, due to Adam historicly being known as a prophet of the Moon. After a great rift between the gods over the fate of humans, the Anunnaki are believed to have left the Moon and returned to their home, the mysterious planet Nibiru**.** In more recent times, some researchers have developed the idea that there are still Anunnaki dwelling in the interior of the Moon, all the while observing mankind's progress. Researcher and author C.L. Turnage states in her book *ET's Are on the Moon & Mars* (page 7) that these beings "were actually present during the time our astronauts made their historic Lunar landings."

Apollo

The ancient Greek god of the Sun. NASA's **Apollo program** was named after Apollo. The name was inspired by images of Apollo on his chariot traversing the skies. The program was given its name by **Abe Silverstein**, an American engineer who was the former Director of Space Flight Development at NASA.

Apollo 1

The first crewed mission of the **Apollo program**. Its official name was *AS-204.* It was to be the first low Earth orbital test of the Apollo

Command/Service Module with a crew. It was scheduled to launch on February 21, 1967. Crew members included Roger B. Chaffee, Virgil "Gus" Grissom and Ed White. The mission never took place. On January 27, 1967, during a pre-flight test, a flash fire quickly spread through the *Apollo* one command module. All three astronauts inside perished.

Apollo 2 through 6

There were no Apollo Space missions specifically named *Apollo 2* through *6*. There were missions that were unmanned and assigned to test different aspects of the **Apollo program**. ***Apollo*** *7* was the first manned mission after ***Apollo 1***.

Apollo 7

Apollo 7 launched on October 11, 1968. The crew consisted of Commander Walter Schirra Jr.; Lunar Module Pilot R. Walter Cunningham; and Command Module Pilot Donn F. Eisele. It was the first **Apollo** mission to transport **astronauts** into space, with the goal of eventually placing men on the lunar **surface**. Reportedly, the *Apollo 7* crew took photographs that showed a large unidentified flying object that has been variously described as "angelic," "beautiful," "large" and metallic." According to the story, the astronauts attempted to cover up the **UFO** shown in the photographs with duct tape. Oddly, on one picture the UFO can still be viewed, while two others show that images have been concealed with the tape. It is speculated that whatever the mysterious object was, it was made by highly intelligent beings that are far more advanced than mankind. UFO proponents hold that the Earth's space programs are being watched and that **extraterrestrials** are especially interested in man's progress into space.

Apollo 8

Apollo 8 launched on December 21, 1968. Its goal was lunar **orbit**. The **astronauts** worked in both technical and scientific areas. The crew consisted of Commander **Frank Borman**; Lunar Module Pilot William Anders; and Command Module Pilot **James Lovell**. It was during this mission that Anders took the now famous "Earthrise" photograph on December 24, 1968. During the *Apollo 8* mission, strange, unexplained events occurred. On their way to the Moon, the **astronauts** were astonished to witness a fast-moving, circular shaped **UFO** that emanated a light that was so bright and powerful, they could barely see within the craft. As the UFO passed, their capsule pitched and yawed, causing the astronauts to nearly lose control of the vehicle. At that point, Lovell made the comment to mission control, "We have been informed that Santa Claus does exist." The term "Santa Claus" some speculate was a code name used by the astronauts for either a UFO or **extraterrestrial** vessel. At one point the **astronauts** heard strange, garbled sounds coming from over the radio. There was no explanation as to what the perplexing noises meant or where they originated. Additionally, there was a peculiar high frequency radio noise that the astronauts described as being "intolerable."

Apollo 9

Apollo 9 commenced on March 3, 1969. It was the **Apollo program's** third crewed mission. It was tasked with performing key procedures for landing men on the Moon and testing the Lunar Module in space for the first time. The Lunar Module would eventually take **astronauts** of the future to the **surface** of the Moon and later return them to the awaiting Command Module in space. The crew consisted of Commander **James McDivitt**; Command Module Pilot **David Scott**; and Lunar Module Pilot Rusty Schweickart. On March 10, 1969, the crew had an unusual sighting. There were large, mysterious, **cylindrical-shaped** objects crossing the moon. The astronauts photographed them. To date, there has been no explanation given as to what the objects were. There have been several reports in **UFO** circles of cylindrical-shaped objects seen in the skies over Earth. Ufologists speculate that what the astronauts took picture of traversing the Moon that day was **extraterrestrial** related. Others maintain that they were crosshairs on the camera.

Apollo 10

The last **Apollo** mission before the history making ***Apollo 11*** that placed men on the lunar **surface**. It was the **Apollo program's** fourth crewed mission and the second time Apollo **astronauts** would orbit the Moon. The crew included Commander Thomas Stafford; Lunar Module Pilot **Eugene Cernan;** and Command Module Pilot John Young. As they traveled to the Moon, the *Apollo 10* crew experienced a problem when the Lunar Module first separated from the Command Module. It was the Lunar Module that would eventually carry two astronauts on the *Apollo 11* mission to the Moon's surface and *Apollo 10* was tasked with performing a practice session for the first piloted landing on the Moon. As the Lunar Module approached the lunar surface, the crew needed to get a lock on the signal from the Command Module, which would enable them to maintain a precise altitude from the lunar surface. They were unable to achieve this goal, due to a transponder failure on the Command Module. Without a lock, the entire mission was at risk. Consequently, the crew in the Lunar Module became stuck in space. The astronauts attempted manual override and failed. They contacted ground control for guidance with no success. During the commotion, Cernan (the Lunar Module Pilot) noticed something moving outside and near the craft. The object then quickly vanished. Minutes later, the circumstances within the Lunar Module improved. The radar systems started working and the crew was able to secure a lock. Mission control then gave them the go ahead to continue towards the lunar **surface**. (*Apollo 10* was a dress rehearsal for *Apollo 11* and stopped just short of landing on the Moon).

When the astronauts returned home, a video that Cernan had taken of the mysterious **UFO** was reviewed. There were mixed thoughts among officials as to what exactly the astronauts had seen. Many believed that what had occurred in space was purely coincidence. Others speculated that the UFO appearing at that time was no accident. Some believed the UFO to be a ship belonging to **extraterrestrials** attempting to warn the astronauts away by disrupting their mission. Others speculate that it was benevolent extraterrestrial beings helping the crew by mysteriously repairing the stalled equipment. Another possibility is that the UFO was emitting signals that were interfering with the radar system on *Apollo 10*. The nature of the UFO and its relation to the troubles of *Apollo 10* are still considered a mystery to this day.

Apollo 11

The historic **Apollo program** mission that placed men on the Moon. *Apollo 11* launched

Apollo 11 mission patch.

on July 16, 1969. The **astronauts** were charged with exploring the Moon and gathering samples for scientific study. Their destination was the **Sea of Tranquility**. The crew consisted of Commander **Neil Armstrong**; Command Module Pilot Michael Collins; and Lunar Module Pilot **Buzz Aldrin**. It was hoped that if *Apollo 11* was successful, there would be more missions of its kind in the future, with the possibility of a lunar **colony** one day being placed there. Armstrong and Aldrin spent eleven hours on the lunar **surface**. There are mysterious and sometimes frightening tales surrounding the alleged experiences of the *Apollo 11* astronauts. Some have nearly become legend among ufologists and conspiracy theorists. On their journey to the Moon, the crew encountered a **UFO** that was trailing them. In his first autobiography, titled *Return to Earth,* Aldrin wrote about a time he was in the *Columbia* (Command Module) looking out into space, when he noticed a strange light outside the craft. In his book he described it as being "brighter than any star," and said that it looked like an "illuminated L." The brilliant light was following the Columbia. The astronauts contacted mission control, but out of caution did not tell the ground personnel that they were being followed by a UFO. Instead, they inquired as to the whereabouts of the *Saturn V* launch rocket. They were informed that the *Saturn V* was around six thousand miles away from their position. Researchers speculate that what the astronauts saw that day was a part of the *Saturn V* or some other piece of equipment. Others maintain that it was quite possibly an extraterrestrial spacecraft.

Apollo 11: Neil Armstrong, Michael Collins and Buzz Aldrin.

In another tale, Armstrong and Aldrin purportedly conveyed a message to mission control, in code, that there were extraterrestrial **spaceships** parked on the rim of a **crater** and that they were being watched by whoever or whatever were in the ships. According to the story, when reporting their discovery to mission control, there was a two-minute radio silence so that the public would not hear the message. The **extraterrestrial** ships are said to have left shortly after the **astronauts** stepped onto the lunar surface. In his book *Our Cosmic Ancestors*, former NASA Communications Engineer **Maurice Chatelain** writes, "Moments before Armstrong stepped down the ladder to set foot on the Moon two UFOs hovered overhead."

In his book *Visitors from Other Worlds* (page 56), author and researcher Brad Steiger writes about Dr. Sergei Bozhich who witnessed the Russians observing the *Apollo 11* moon landing. Steiger writes, "In his [Bozhich's] opinion, the two UFOs appeared ready to assist the US astronauts in case anything should go wrong with the landing. Once the module appeared to be securely settled on the lunar surface, the alien spacecraft flew away."

There is a tale that has circulated for years stating that when one of the astronauts opened the door of the Lunar Module, he immediately saw an extraterrestrial. The being is said to have had an ethereal appearance. Additionally, as the astronauts walked on the Moon, It is said that there was an unexplained bright light emanating from a nearby crater. See ***Apollo 11* Goodwill Messages.**

Apollo 11 Goodwill Messages

Messages of hope and congratulations from world leaders of 73 countries to the

United States and the ***Apollo 11*** team. The messages were placed on a small silicon disc, approximately the size of a half dollar. The disc was placed inside an aluminum container, then a pouch and was left on the Moon by the *Apollo 11* **astronauts** at the **Sea of Tranquility** along with other articles. Some of the messages are handwritten, others typed and some are in the countries' own language. The disc also includes messages from Presidents Eisenhower, Kennedy, Johnson and Nixon. Additionally, there are the names of the United States congressional leaders of the day, the names of members of the four committees of the House and Senate that were responsible for **National Aeronautics and Space Administration** legislation and the names of NASA's upper management staff. At the top of the disk there is an inscription that states: "Goodwill messages from around the world brought to the Moon by the **astronauts** of *Apollo 11*." A second inscription can be found along the edge of the disc. It states: "From Planet Earth –July 1969."

Apollo 12

The first mission after the history-making ***Apollo 11*** and the second to land on the lunar **surface**. Its destination was the Ocean of Storms. The team included Commander **Charles Conrad Jr.,** Command Module Pilot **Dick Gordon** and Lunar Module Pilot **Alan Bean**. *Apollo 12* was fraught with incomprehensible and baffling events. In his book *Our Mysterious Spaceship Moon,* author and researcher Don Wilson writes, "There is widespread agreement by researchers (unquestionably backed up by authenticated evidence from NASA files) that mysterious, unexplainable things did happen on the expedition of *Apollo 12*." The first incident occurred thirty seconds after launch when a lightning bolt hit the *Saturn V* rocket. Thirty seconds later, a second bolt hit it. All power was lost for three minutes. The bolts of lightning could not be explained. Even though there was rain on that day, the nearest thunder and lightning storm (per weather reports at the time) was approximately twenty miles out. For a few minutes, there was a question as to whether *Apollo 12* would continue or abort the mission. Even more mysteriously, reports came in from observatories from around Europe stating that there were two bright objects with flashing lights pacing the *Saturn V*. The following day, the crew communicated to NASA that they too had seen two **UFOs** following their spacecraft. One of the objects was reportedly revolving as it flew. Eventually both UFOs took off at a high rate of speed and vanished. The **astronauts** stated later that they believed that whoever was operating the UFOs were benevolent. Ufologists speculate that **extraterrestrials** were watching the launch and were there to observe the mission. Others believe that the lightning bolts were purposely sent as a warning.

Aldrin jumping on the Moon during the *Apollo 11* mission.

While in space, the crew and mission control heard inexplicable, baffling sounds. The source could not be explained or located. On the return to Earth, while in close proximity of the planet, *Apollo 12* again reported seeing a UFO that the astronauts described as having a bright red, pulsating light. It eventually disappeared. No credible explanations were ever given for any of the mysterious sightings and unexplained noises that were linked to *Apollo 12*.

A photograph taken during the *Apollo 12* mission and one that has received attention from ufologists, is NASA photograph *AS12-497319*. In the picture, there appears to be a large UFO floating above one of the astronauts as he is walking on the Moon. Additionally, as the crew went about their work on the lunar surface, they saw a strange, semi-translucent object. They described it in detail stating that it was shaped like a **pyramid** and radiated the colors of the rainbow. It was hovering just above the lunar floor. The astronauts felt that they were being closely monitored by the strange object.

Apollo 13

The seventh **Apollo program** mission to the Moon. *Apollo 13* launched on April 11, 1970. It was slated to be the third mission to land on the moon. The crew was made up of Commander **James A. Lovell Jr**.; Command Module Pilot John L. Swigert Jr.; and Lunar Module Pilot Fred W. Haise Jr. Due to a break in the oxygen tank on the Service Module, there was an explosion on board the craft, leaving *Apollo 13* crippled. Unable to land, it orbited around the Moon instead. Subsequently, *Apollo 13* returned to Earth without ever completing its mission. Some believe that **extraterrestrials** assisted *Apollo 13* in returning home. According to the story, the crew allegedly received a communication from an extraterrestrial spacecraft, instructing them on how to safely return to Earth. It is said that without the assistance of these beings, the **astronauts** would not have made it home. Some claim that when they returned to Earth, the crew was told not to reveal the truth about what really happened in space.

Apollo 14

Apollo 14 was the replacement for the ill-fated ***Apollo 13***. *Apollo 14* launched on January 31, 1971. Its destination was the Fra Mauro **crater**. Scientists sought information from the area, as they believed it to be a site that held historic data that would assist them in gaining more knowledge about the Moon's past. It included crew members Commander **Alan Shepard**; Command Module Pilot Stuart A. Roosa; and Lunar Module Pilot **Edgar Mitchell**. While on the Moon, the astronauts took pictures that revealed a number of mysterious images, including strange **blue lights** and anomalous objects.

Apollo 15

The fourth **Apollo program** mission to land on the Moon. It was the first of the "J missions," and the first time that **astronauts** used the Lunar Roving Vehicle on the Moon's **surface**. *Apollo 15* launched on July 26, 1971. It was more scientifically oriented than **Apollo** missions in the past. The crew included Commander **David R. Scott;** Command Module Pilot **Alfred Worden**; and Lunar Module Pilot **James Irwin**. The Lunar Module crew landed in the Hadley-Apennine region, a dangerously rugged area, filled with crater-pocked terrain, **rocks** and boulders. Due to the scientific goals of the trip, the Command Module was given the name *Endeavour,* after Captain Cook's vessel from his landmark science expedition in 1768. As a symbol to honor Cook, the *Apollo 15* crew carried a piece of wood from the sternpost of the original *Endeavour*. A number of strange incidents and tales are associated with *Apollo 15*. One account states that astronauts David Scott and James Irwin, while working on the surface of the Moon, were nearly struck by a mysterious **object** that streaked across the lunar sky. It was unclear what the object was or where it came from. The *Apollo 15* astronauts returned to earth with 170 pounds of sample material removed from the Moon. This included their most important discovery of the mission, the **Genesis Rock,** which turned out to be approximately 4.5 billion years old.

Apollo 16

The fifth **Apollo program** mission to land on the Moon. *Apollo 16* was launched on April 16, 1972. Its crew members included Commander John Young; Command Module Pilot **Thomas Kenneth Mattingly;** and Lunar Module Pilot **Charles M. Duke Jr.** Their goal was the Descartes Highlands on the Moon's nearside, an area of treacherous terrain encompassing mountains, **craters** and odd white and pink shimmering boulders. During the *Apollo 16* mission the **astronauts** gathered soil samples, took photographs and conducted experiments. The *Apollo 16* astronauts reportedly witnessed mysterious flashes of light while they were in **orbit**. The source of the lights could not be located. On another occasion, as Ken Mattingly went about his duties on board the Command Module, a weird light suddenly appeared and streaked quickly across the horizon. It disappeared behind the Moon's **far side**. The astronauts' stories were perplexing to NASA scientists. Several explanations were offered. One theory was that a micrometeorite hit the Moon. Some held that the lights were the result of cosmic rays affecting the men's eyesight. These hypotheses were eventually dismissed. It was concluded that there was no explanation for what the crew had seen. See **Glass-Crystal Structure.**

Apollo 17

The **Apollo Program**'s last mission to the Moon. *Apollo 17* was launched on the night of December 7, 1972. The astronauts' goal was to land in the Moon's **Taurus-Littrow** valley. The crew included Commander **Eugene Cernan**; Command Module Pilot **Ronald Evans;** and Lunar Module Pilot **Harrison Schmitt**. Their assignment was to survey the Moon, photograph areas of interest and perform special tests on equipment. There is a strange tale that states that the *Apollo 17* **astronauts** found a dead **cosmonaut** on the way to the Moon. Purportedly, once *Apollo 17* neared the Moon, the crew saw flashes of light emanating from the vicinity of the Moon into their craft. Bewildered, the astronauts attempted to locate the source of the light, but were unsuccessful. Instead, they found a deceased cosmonaut hovering outside of their spacecraft.

The astronauts of *Apollo 17* took photographs that had anomalous lights and unusual objects. In one photograph in particular (NASA image AS17/AS17-151-23127), there is a peculiar illuminated object located near the lunar floor. Some **UFO** researchers have likened it to a space portal. Some refer to it as a Stargate (a doorway to other worlds). In an article titled "Did Apollo 17 find a Stargate on the Moon," researcher and author **Michael Salla** writes, "The object appears to be a space portal of some kind with an eerie blue glowing ring around a central darker portion." There is no more explanation as to what the curious anomaly may be. Additionally, the *Apollo 17* astronauts located very small, spherical orange glass beads inside of the Shorty **crater**. In the book *Elder Gods of Antiquity* (page 102), author M. Don Schorn writes, "The analysis of those beads dated the material at 3.8 billion years old. Nowhere else on the moon has such a comparable material yet been found or observed."

In a video titled *Alien Moon Structures–Apollo 17* that was researched and scripted by Benjamin Redman, there is a report about an incident that occurred in 1972 as astronauts **Eugene Cernan** and **Harrison Schmitt** were exploring the Moon. As the two were traversing the lunar **surface** in the Lunar Rover, traveling through the **Taurus-Littrow** valley, NASA was controlling the cameras from ground control, as the mission was being televised. Suddenly, there appeared a strange sight in front of the cameras, something that looked like a massive boxed-shaped structure. Reportedly, broadcaster **Walter Cronkite** who was covering the mission

for *CBS News* stated excitedly, "that looks like a man-made structure!" Quickly, the feed was cut and footage that was shot earlier in the day replaced it. Millions of viewers are said to have witnessed what was later perceived as a blunder with the camera. Purportedly, what the astronauts and also the viewers had seen was a massive rectangular structure that has been described as windowless and resembling a shoebox. The reason given to viewers for the twenty-minute delay, was that the camera on the Lunar Rover had inadvertently taken a photograph of itself. There are some that do not believe this story. It is believed that the astronauts did encounter a large structure on the Moon that day and that it was hidden from the public. There are those who believe that some comments made by Eugene Cernan about the Moon were a cryptic message about their seeing an extraterrestrial structure while there. Cernan stated, "I'm on the surface and, as I take man's last step from the surface, back home for some time to come—but we believe not too long into the future—I'd like to just [say] what I believe history will record. That America's challenge of today has forged man's destiny of tomorrow. And, as we leave the Moon at **Taurus-Littrow**, we leave as we came and, God willing, as we shall return: with peace and hope for all mankind. Godspeed the crew of *Apollo 17*." There is an in-depth analysis of the witnessing of the box-shaped structure in **Taurus-Littrow** at *Alien Moon Structures–Apollo 17,* https://www.youtube.com/watch?v=lVyaWOlnuo0.

Apollo 18
(Movie)

A movie based on alternate history and what would have been the **Apollo program's** last mission to the Moon. The movie's advertisement states, "Decades-old found footage from NASA's abandoned *Apollo 18* mission, where two American **astronauts** were sent on a secret expedition, reveals the reason the US has never returned to the Moon." The film stars Warren Christie, Lloyd Owen and Ryan Robbins. The director was Gonzalo Lopez-Gallego. Screenplay writers included Brian Miller and Cory Goodman. The release date was September 2, 2011.

Apollo 18 through 20

Apollo 18–20 were **Apollo program** missions that were said to have been in the planning stages, but never made it to fruition. The missions were being planned after *Apollo 17* (the last **Apollo** mission to the Moon). In 1972 they were cancelled. It has been speculated that there are reasons that the program was cancelled beyond what the public has been told. Conspiracy theorists believe that there are secrets about the Moon that are being kept from the general public. It has been rumored for years that Apollo **astronauts** were warned off the **Moon** by **extraterrestrials.** The government's official reason for cancelling the **Apollo program** was that it was due to budget cuts and a lack of public interest.

Apollo 20 Hoax

A faux story that led people to believe that there had been a clandestine ***Apollo 20*** mission to the Moon. According to the tale, the mission took place in August 1976. It was sent to explore a mysterious cigar-shaped **object** located near the Delporte crater on the **far side** of the Moon. The object had been seen in NASA photographs taken during the ***Apollo 15*** mission. After studying the photographs, officials believed it to be an ancient extraterrestrial spaceship. It was decided to send a secret mission to the Moon to investigate it. **Apollo 20** (a combined mission between the United States and Russian reaches the **far side** of the Moon, where the **astronauts** find remnants of a city and as predicted, an enormous, ancient alien spacecraft. The astronauts board the craft and discover the body of an **extraterrestrial** humanoid female that looked to be the ship's pilot. She appeared to be in stasis. Apparatuses were attached to her eyes and nose that looked as if they were used for flying the spaceship. Film footage was taken of the expedition by one of the astronauts. They later returned to Earth with the body of the extraterrestrial.

In April 2007 several videos were posted on YouTube under the name retiredafb. The person posting the footage was named **William Rutledge.** Rutledge claimed to have been born in Belgium in 1930. He was a US citizen, retired and living in Rwanda. Rutledge claimed to have been one of the astronauts from the mission (his comrades were Leona Marietta Snyder and Alexei Leonov). The videos show a city on the **far side** of the Moon, a cigar-shaped craft and pictures of the extraterrestrial pilot whom they named **Mona Lisa**. A French sculptor by the name of Thierry Speth eventually came forward and claimed responsibility for creating the story and artwork in the video.

Apollo Astronauts
See **Astronauts**.

Apollo Astronaut Transcripts
See **Appendix 1: NASA Transcripts**.

Apollo Mission(s)
See ***Apollo 7; Apollo 8; Apollo 9; Apollo 10; Apollo 11; Apollo 12; Apollo 13; Apollo 14; Apollo 15; Apollo 16; Apollo 17;*** **Apollo Program.**

Apollo Program
(Project Apollo)

The Apollo Program was the third spaceflight program or the **National Aeronautics and Space Administration** (NASA). The program ran from 1969 to 1972, resulting in **astronauts** landing and walking on the Moon. The Apollo program was named after the Greek god **Apollo** who is often depicted on a chariot traversing the heavens and moving past the Sun. The goal of the Apollo program was to place men on the Moon. The United States and the **Soviet Union** were competing to be the first to have men land on the lunar **surface**. President John F. Kennedy wanted to accomplish this within the time frame of the 1960s. On July 20, 1969, **Neil Armstrong** and **Buzz Aldrin** of ***Apollo 11*** made history when they landed on the Moon and walked on the **surface**. In total, six Apollo missions landed on the Moon and twelve **astronauts** walked and worked on the lunar **surface**. The astronauts explored parts of the Moon, took photographs, ran scientific tests and brought back samples.

There are theories that revolve around the reasons for the missions to the Moon. One is that government officials believed that an ancient civilization had once existed there and were interested in investigating and confirming that possibility. Another is that officials knew that there was something or someone on the Moon due to various reports of **anomalous lights** and objects resembling constructions appearing and disappearing. Another theory is that some world governments sought to eventually put a **colony** on the Moon and therefore wanted firsthand knowledge of the Moon's make up. Many believe that instead of scouting sites for a human colony, those sent to the Moon found ancient ruins of a past civilization. Conspiracy theorists believe that something significant happened while astronauts were on the Moon and that is the reason for the Apollo program abruptly coming to an end. The official story is that the Apollo missions were cancelled due to the exorbitant cost of the program. See **Apollo 7; Apollo 8; Apollo 9; Apollo 10; Apollo 11; Apollo 12; Apollo 13; Apollo 14; Apollo 15; Apollo 16; Apollo 17.**

Apollonius of Rhodes
295 BC – Unknown

A prominent ancient Greek author best known for his poem *Argonautica*, in which he tells the story of Jason and the Argonauts. Apollonius spoke of a period on Earth when there was no Moon in the sky. It was a time, he says, before the Danaans and Deucalion races appeared. In *Argonautica IV. 264,* he writes, "when not all the orbs were yet in the heavens, before the Danaans and Deucalion races came into existence and only the **Arcadians** lived, of whom it is said that they dwelt on mountains and fed on acorns, before there was a moon." His statement has been used in the argument suggesting that there was a time when the Moon didn't exist and that it was brought into Earth's orbit by some mysterious means. See **Pre-Lunar Earth**.

Arcadians

A race of people that lived in ancient Greece, known also as the **Pelasgians** and **Proselenes.** Their civilization is believed to have been older than the Moon. The ancient poet **Ovid** wrote that the Arcadians lived on the land "before the birth of Jove" (meaning a time before the worship of the Roman god Jupiter). This period is believed to have been before the civilizations of Mesopotamia and Egypt which were at the cradle of civilization.

Archimedes (Crater)

A prominent impact **crater** located in Mare Imbrium (Sea of Showers). Archimedes stands out due to its smooth bottom and because it is filled with mare basalt. It is a place where strange looking, large **cylindrical-shaped** objects and in one case strange lights have been seen. These objects have been estimated to measure twenty miles long and three miles wide. They are believed by some to be **extraterrestrial** spacecraft. In her book *ET's Are on the Moon & Mars,* C.L. Turnage writes, "A strange collection of gigantic, cylindrical objects can be seen resting on the floor the crater Archimedes." On March 29, 1966, bands of light were seen covering the crater floor. There is no explanation for the odd lights. Scientists believe that Archimedes is simply a crater with unusual features.

Archimedes Platform

A large bottle-shaped formation on the lunar **surface**. In a photograph taken by NASA's *Lunar Orbiter 4* near the **Archimedes crater**, there is what some interpret as a raised structure. In his book *Extraterrestrial Archaeology (*page 137*),* author David Hatcher Childress informs us that the construction (as reported by David Darling) is five miles long and one mile wide, stretching 5,000 feet high. Some scientists consider it to be a natural part of the Moon's scenery, while others believe it to be an artificial structure placed there by otherworldly beings.

Arianrhod

A goddess of the Moon in Celtic beliefs. The name Arianrhod means "silver wheel," denoting the wheel of the year and the cyclical change of seasons.

Aristarchus (Crater)

A large impact **crater** that was formed nearly 450 million years ago. It has been called the "single brightest spot on the Moon." It is twenty-five miles in diameter and is estimated to be twice the depth of Arizona's Grand Canyon. It's named after the legendary Greek astronomer Aristarchus (Aris-TAR-kess) of Samos. Various types of lights and strange illuminations have been reported near the crater. In fact, **Aristarchus** harbors one of the Moon's greatest mysteries, the "domed **blue light**." There appears to be a **dome** radiating a strange brilliant, blue light that illuminates the entire region. This blue light has been observed for over a hundred years, with no explanation for it. It is so spectacular that it can be seen from Earth through normal binoculars. It is said to be especially bright when earthshine is present. As time has progressed and more research has been done near the crater, the findings have led some to speculate that someone or something is producing energy from inside the Moon. **Extraterrestrial** enthusiasts maintain that there is a power source inside the crater that is causing the strange light. They speculate that this source may be operated by beings that are living on the Moon. Others hypothesize that the crater may be a gigantic helium-3 fusion reactor as the Moon is believed to have significant amounts of helium-3. Some scientists maintain that the reason Aristarchus shines so brilliantly is because it is such a young crater. They hold that Aristarchus will eventually darken from the effects of sunlight and solar radiation, a process which occurs with all lunar craters. In addition to the illuminated blue dome, other mysterious lights and anomalous activity have been reported around Aristarchus for centuries. A compilation follows.

1783, March (day unknown), luminous points of lights were seen.

1786, December 24, a brilliant light was witnessed.

1787, May 19-20, an extremely bright light was observed.

1788, April 9, the crater appeared extraordinarily bright for approximately an hour.

1788, December 2, a brilliant light appeared inside the crater at 5:35 am. It was monitored for thirty minutes.

1789, March and April (exact dates unknown), several illuminated spots of lights were reported being near the crater and also its darkened area.

1821, May 4-6, a bright point of light was seen by observers in the vicinity of the dark side the crater.

1822, June 22-23, a **transient lunar phenomenon** was witnessed by an observer who described it as a lunar volcano.

1824, May 1, a blinking light was witnessed near the darkened area the crater.

1824, October 18, there was a sighting of overlapping, various-colored orbs of lights. They were observed in the west and north west areas.

1866, June 10, a shimmering light was seen. On June 14-16, a reddish-yellow light was observed.

1867, April 9, a bright spot of light was monitored for an hour and a half in the vicinity the crater's darkened area.

1867, May 6-7, for several hours each night, an area the crater shined brilliantly. It was described as having a volcano-like appearance.

1867, May 7, A reddish-yellow light, resembling a beacon, was seen in the vicinity the crater.

1881, August 6 and 7, an intense violet light covered the region, giving the appearance of the area being covered with fog.

1881, December 5, during a **lunar eclipse**, the entire crater was illuminated.

1897, September 21, glimmering streaks of light could be seen.

1898, December 27, during a lunar eclipse, the crater was described as shining brilliantly.

1903, March one and March 3, intermittent lights resembling little stars were reported.

1905, February 19, during a lunar eclipse, an illuminated starlike spot could be seen.

1937, September 17, a bright streak of light was observed.

1939, February 23, an area was found to be glowing.

1940, December 2, the crater shined brightly and could easily be distinguished in the Moon's darkened hemisphere.

1949, November 3, a blue glare was seen at the base inner west wall.

1950, June 28, a blue light was seen at the rim of the west wall. On June 29 a blue light was observed on the east wall.

1954, July 14, July 16, July 17 and July 24, a violet colored light was seen in various areas.

1954, August 11, a brilliant light was observed.

1955, January 8, January 12, April five and May 5, a violet to blue-violet light was seen in various areas. The light varied in intensity in brightness.

1955, August 3, in the plateau area, a faint violet light was observed. On August 30, a pale blue light was seen in areas around the crater.

1955, September 7, September 8 and September 9, a brilliant strong violet-blue light was seen in various areas around the crater.

1955, October 2, the crater appeared obscure in some areas due to a misty violet colored haze.

1955, October 4, a pale violet glare was seen. On October five there was a brilliant blue-violet light. On October 31, there were various intensities of blue light around the crater.

1956, November 17 to 18, an intense illuminated light was witnessed.

1957, March 17 and March 18, an intense violet light was observed.

1957, June 11, a blue glow was viewed covering the floor.

1957, August 18, a soft blue light glowed on all of the walls, while the floor shimmered in white.

1957, October 12, a brilliant streak of light appeared. Afterward, a strange brown color followed.

1958, May 1, the sunlit portion of the floor, had a blue tint to it.

1958, May 1, May 4, May 31, June 29, July 2,1959, September 5, a spot of light described as being colorless and starlike was observed.

1961, November 26, November 28 and December 3, a red glowing light appeared.

1963, October 30, reddish-orange to ruby colored areas were seen.

1963, November 11, strange color changes were observed.

1964, June 25, during a lunar eclipse, the crater shined brightly.

1964, June 27, a glowing blue-violet light was seen near the east wall and northeast rim.

1964, October 22, October 24, October 25, November 21, November 23 and November 24, a bright blue-violet light was monitored.

1965, July one and July 2, a starlike image was observed on the dark side.

1965, July 3, a pulsing orb of light on the crater's dark side was monitored for an hour and ten minutes.

1965, September 9, an orange-red strip was discovered on the floor.

1965, April 2-3, the central peak shined brightly.

1966, May 1, an area glowed in a mysterious red color for fifty minutes. A small luminous orb of light shined on the northwest wall.

1967, April 22, the crater shined so brightly, that it could be observed with the naked eye.

The sightings of strange phenomena in the crater continues to this day.

Aristotle
384 BC–322 BC

Prominent ancient Greek philosopher and scientist. Aristotle spent much of his sixty-two years researching, instructing and writing about cosmology. He taught that our world is positioned in the middle of the cosmos and that the Moon is a perfectly formed sphere that revolves around it. In explaining the dark regions of the Moon, he maintained that these areas assimilate and radiate light dissimilarly from other areas of the Moon. He believed that the movement of the Moon is generated through a special force, a prime mover that controls all events in the universe. To honor Aristotle, the International Astronomical Union named an impact **crater** after him in 1935. Aristotle spoke about a mysterious race of people that existed before there was a moon in the sky. His writing about this race of people lends credence to the idea proposed by some researchers that the Moon is not a natural **satellite**. In his work *Constitution of Tagean,* he tells of a people, referred to as the Pelasgians, who inhabited Arcadia before the Greeks. He wrote that they held claim to the land first since they were, as he stated, "already living on it before there was a moon in the heavens." See **Aristotle Crater**.

Aristotle (Crater)
(Aristoteles)

An impact **crater** that is named after the renowned Greek philosopher **Aristotle**. The name is officially "Aristoteles," the conventional form of Aristotle. The crater is approximately 48 miles in diameter and nearly 10,000 feet deep. In 1882 in the Aristotle region of the Moon's **surface** a number of unexplained dark shadows could be seen. Mysteriously, the source of these shadows could not be explained.

Arkhipov, Alexey V.
1959–

Arkhipov is a Ukrainian radio astronomer and astrophysicist. He has done a great deal of work in the area of SETA (Search for **Extraterrestrial** Artifacts). He particularly focuses on the Moon and the Earth. Arkhipov established Project SAAM (Search for Alien Artifacts on the Moon) in 1992. Says Arkhipov, "Our Moon is a potential indicator of a possible alien presence near the Earth at some time during the past 4 billion years."

Armstrong, Neil Alden
1930–2012

A former Apollo **astronaut**, best known for being the first man to walk on the Moon. Armstrong was a part of the *Gemini 8* mission, where he served as the command Pilot and was the commander of the historic ***Apollo 11*** mission to the Moon. When Armstrong took his first step onto the lunar **surface** he stated, "That's one small step for man, one giant leap for mankind." The audience for this momentous event was international; it was watched by millions of people around the world. When the Lunar Module set down on the Moon's surface Armstrong stated, "The *Eagle* has landed," to let the world know that the trip to the lunar surface was a success.

There are several mysterious stories that have been circulated about Neil Armstrong and his experiences while on the Moon. There is a tale that states that Armstrong and his comrade, astronaut **Buzz Aldrin**, witnessed **extraterrestrial** spacecraft sitting on the rim of a **crater** after exiting the Lunar Module and stepping onto the Moon's surface. As the story goes, Armstrong radioed mission control about what he and Alrdrin were seeing. According to the account, there was a two-minute delay so that the public could not hear what was being said. Shortly afterward, Armstrong allegedly switched to a medical channel to keep the conversation private. Even though these precautions were taken, the conversation was said to have been picked up by people with VHF receiving facilities, who later relayed what they had heard, which was Armstrong describing seeing large **spaceships** and saying that they were being watched. Another account states that two **UFO**s were hovering above Armstrong and Aldrin as they stepped onto the Moon's surface.

Vladimir Azhazha, a former president of the USSR Academy of Science and a prominent ufologist has been outspoken on alleged mysteries surrounding UFOs and space mission cover-ups for years. His ideas, thoughts and comments have become well-known. In *Above Top Secret: The Worldwide UFO Cover-up* by Timothy Good (page 384), Azhazha is quoted as saying, "Neil Armstrong relayed the message to mission control that two large, mysterious **objects** were watching them after having landed near the moon module. But his message was never heard by the public because NASA censored it." It has been said that after the *Apollo 11* mission, Armstrong led a very private life, rarely giving interviews or talking about his Moon experience. Some speculate that Armstrong witnessed something so mysterious while on the lunar surface that he spent the rest of his life quietly contemplating the events on the Moon.

Neil Armstrong as a test pilot in 1956.

One conspiracy theory suggests that Armstrong's silence proves that there was never any Moon landing. It is said that because he was a man of high morals and integrity, he simply did not want to perpetuate a lie. Another hypothesis suggests that Armstrong was quiet about his Moon mission because he saw extraterrestrials on the Moon and was instructed not to relay what he had seen. In 1994, during a White House ceremony celebrating the twenty-fifth anniversary of the first Moon landing, Armstrong gave a speech. Some believe that statements in his speech had an underlying meaning alluding to there being mysteries on the Moon. Armstrong stated, "Today we have with us a group of students, among America's best. To you we say, we have only completed a beginning. We leave you much that is undone. There are great ideas undiscovered, breakthroughs available to those who can remove one of truth's protective layers." Armstrong died on August 25, 2012 from heart failure. Many believe that he took secrets about the Moon with him when he died.

Artemis

Moon goddess in Greek mythology. Artemis rules the night sky. Her symbol is the **crescent moon**, which she wears upon her head.

Artifact(s)

See **Asada (Crater); Lobachevsky (Crater);** ***Lunar Reconnaissance Orbiter*****; Modern Day**

Mysteries; Pyramids; Ruins; Tower of Babel; War.

Artificial Moon Theory
See **Spaceship Moon Theory.**

Asada (Crater)

A small impact **crater**. It has been described as having the characteristics of a satellite dish, complete with a circular base and a long rod protruding out of the center. **UFO** researchers argue that this may be either an ancient satellite dish on the Moon, or a relic resembling a satellite dish from an ancient civilization that once existed on the Moon. Some lunar researchers maintain that any resemblance to a satellite dish is due to an illusion created by the various angles from which the Sun hits the crater.

Asimov, Isaac
1920–1992

A Russian-born, American author and the world's most prolific writer. Once a professor of biochemistry at Boston University, Asimov wrote and edited over five hundred publications. His books have been distributed in nine classifications of the Dewey Decimal System. He specialized in science and was a popular science fiction writer. It was Asimov that developed the "Three Laws of Robotics." Asimov believed that the Moon is too large to have been caught in Earth's orbit. He also held that it was too far away. In 1963, in his book *Asimov on Astronomy* (page 124) Asimov wrote: "What in blazes is our moon doing way out there? It's too far out to be a true satellite of earth, it's too big to have been captured by the earth. The chances of such a capture having been effected and the moon then having taken up a nearly circular orbit about the earth are too small to make such an eventuality credible. But, then, if the moon is neither a true satellite of the earth nor a captured one, what is it?" His comments are frequently used by proponents of the Artificial Moon Theory.

Association of Lunar and Planetary Observers (ALPO)

A worldwide organization dedicated to the study of the Moon and other cosmic bodies. The objective of ALPO is to encourage, promote and advance the knowledge of planetary entities through the use of equipment and processes that are available from professional as well as amateur astronomers. Observations from ALPO are published in the periodical *The Strolling Astronomer* (also known as the *Journal ALPO*). Further information about ALPO can be located at http://www.alpo-astronomy.org.

Asteroid Theory

A theory that was once put forward by scientists to resolve the mystery of how the Moon was created. The hypothesis suggests that the Moon may be an asteroid that was captured in Earth's gravitational pull. The idea has since been dismissed.

Astronaut(s)

A person that is trained to travel into outer space. In the United States they are members of the **National Aeronautics and Space Administration** (NASA) that serve in space. Rumors that astronauts experienced mysterious events while in space have been around since the Project Mercury missions (Project Mercury was a prelude to the Apollo program that would take **astronauts** to the Moon). **Scott Carpenter**, one of the astronauts chosen to be a part of the Project Mercury program, once cryptically commented, "At no time, when the astronauts were in space were they alone: there was a constant surveillance by **UFO**s."

There are many tales surrounding the astronauts and their strange and perplexing experiences during the lunar missions. There are stories of the astronauts being followed by unexplained lights and UFOs, seeing baffling

structures and objects on the **surface** of the Moon and witnessing strange **anomalies**. There is one rumor of an **extraterrestrial** sighting. This information has allegedly come from astronauts, Apollo transcripts and people that worked on the **Apollo program**. In his book *We Discovered Alien Bases on the Moon II*, author Fred Steckling writes about astronauts seeing extraterrestrial vehicles on the Moon during the ***Apollo 11*** Moon landing. He claims that photographs taken of UFOs and other **anomalies** from the Moon missions were airbrushed out before being released to the public.

Many feel that mysterious experiences and sightings of astronauts in space are believable due to their training, experience and credibility. The majority of the **Apollo** astronauts have reported experiencing something anomalous in space, seeing strange lights, UFOs or even hearing talking in what seemed to be an otherworldly language coming from over the radio. **See Buzz Aldrin; Apollo 10; Apollo 11; Apollo 12; Apollo 13; Apollo 14; Apollo 16; Apollo 17; Neil Armstrong; Alan Bean; Frank Borman; Eugene Cernan; Charles Conrad Jr.; Charles M. Duke, Jr.; Dick Gordon; James Lovell; Thomas Kenneth Mattingly; James McDivitt; David Scott.**

Astronomer(s)

A scientist that examines the natural objects (i.e. the Sun, Moon, Planets and Stars) located in the cosmos. See **Alexey V. Arkhipov; William R. Brooks; Cecil Maxwell Cade; Winifred Satwell Cameron; Gian Domenico Cassini; Frank Dennet; Nicolas Camille Flammarion; Walter Goodacre; James Greenacre; Franz von Paula Gruithuisen; Edmond Halley; Frank Halstead; John Herschel; William Herschel; Johannes Hevelius; Howard Hill; Hans Hoerbiger; Pierre-Simon Laplace; Rudolph Lippert; Joseph Johann von Littrow; Nevil Maskelyne; Patrick Moore; Nicholas of Cusa; William Henry Pickering; Johann Schroeter; Eugene Shoemaker; Hugh Percy Wilkins; Joseph Zentmayer.**

Atlantis

An ancient legendary island-kingdom first mentioned in the writings of Plato. It is believed to have been located in the Atlantic Ocean, west of Gibraltar. Plato wrote that it sank into the ocean. Some believe it to have been a mythical place, while others contend that it was a real and powerful ancient empire, whose territory mysteriously reached to the Moon. Some researchers maintain that the ancient ruins purportedly found on the Moon by the **astronauts**, may have belonged to a city that was once a part of the Atlantis Empire. The city is believed to have been destroyed in a war. Atlanteans are thought to have had technology that far surpassed what is found in the modern world. Legendary psychic Edgar Cayce wrote of Atlanteans having aircraft and submarines. Their ability to fly is said to have been due to their advanced knowledge. Using solar and electromagnetic energy, they were able to travel back and forth between the Moon and Earth.

Atmosphere

The gaseous layer that encircles a celestial body. In the *Moon Watcher's Companion* (page 101) from 2002, under "Anatomy," the Moon is said to have "practically" no atmosphere. The controversy over whether the Moon has an atmosphere, even a small one, has been around since May 18, 1787, when two astronomers reported seeing mysterious lightening above the lunar **surface**. Today, there are scientists who believe that the Moon does have a scant atmosphere, but maintain that it would take an artificial environment to live on the Moon. Psychic researcher **Ingo Swann,** who used remote viewing to study the Moon for a secret agency, wrote in his book *Penetration*, that while observing the Moon, he witnessed

extraterrestrials that could breathe without devices. The famed extraterrestrial contactee **Alex Collier**, in a lecture he gave in 1996 titled *Moon and Mars,* indicated that on the **far side** of the Moon, there is an atmosphere. An article from NASA, from April 12, 2013, titled "Is There an Atmosphere on the Moon?" states, "Until recently, most everyone accepted the conventional wisdom that the Moon has virtually no atmosphere. Just as the discovery of **water** on the Moon transformed our textbook knowledge of Earth's nearest celestial neighbor, recent studies confirm that our Moon does indeed have an atmosphere consisting of some unusual gases, including sodium and potassium, which are not found in the atmospheres of Earth, Mars or Venus. It's an infinitesimal amount of air when compared to Earth's atmosphere." Some speculate that the mysterious **domes** seen on the Moon may be part of an artificial base built by otherworldly beings beneath the surface of the Moon, due to the inhospitality of the Moon's surface and lack of a relevant atmosphere.

Azhazha, Vladimir
See **Neil Armstrong**.

B

Backhoe

A piece of excavation machinery. A backhoe has a hinged container that is attached to the end of an extended, jointed limb. On the rim of the **Tycho crater** there is an anomalous formation that resembles a backhoe. It was located in a photograph taken by the ***Clementine*** spacecraft and given the name "the backhoe" by author **Mike Bara**, who featured the image in his book *Ancient Aliens on the Moon* (page 188).

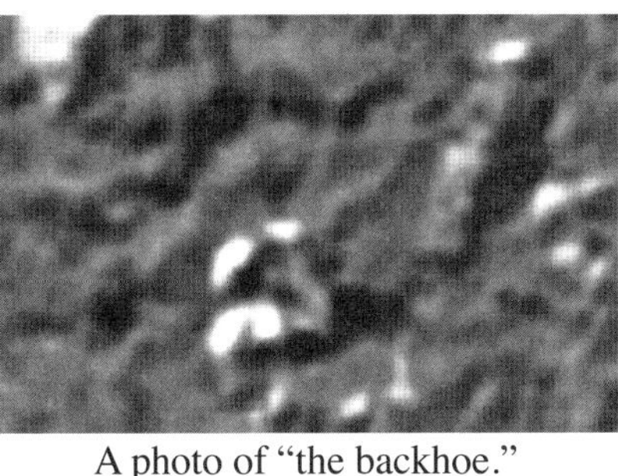
A photo of "the backhoe."

Bara, Mike
(Unknown birth and death dates)

Bestselling author and lecturer. Bara is the author of *Ancient Aliens on the Moon*. He also co-authored *Dark Mission: The Secret History of NASA* with **Richard Hoagland**. In *Ancient Aliens on the Moon,* Bara offers an in-depth look at what he believes are ruins and artifacts located in various photographs from NASA missions. See **Backhoe, Chalet; *Clementine* Spacecraft; Dark Side; Geo-Dome; Longhorn; Pyramids; Tycho Crater.**

Barr, Edward
(Unknown birth and death dates)

A former United States Air Force Aeronautical Chart and Information Center (ACIC) cartographer. On the evening of October 29, 1963, Barr and his associate **James Greenacre** were working at Lowell Observatory in Flagstaff, Arizona. Their job was to map the **Aristarchus crater** which was 27 miles in diameter. Both Barr and Greenacre were familiar with that area of the Moon. That evening, the two were astonished to see what has been described as an intensely glowing, reddish-orange and pink light on the Moon's **surface**, located near Cobra's Head and the southwest interior rim of the **Aristarchus crater**. The reported sighting of these two respected astronomers of this **transient lunar phenomenon** (TLP) gained the attention of others in the field and inadvertently sparked more interest in the observing and the recording of TLPs on the Moon. A report titled "**Lunar Color Phenomena, ACIC Technical Paper No. 12,"** published by the United States Air Force Aeronautical Chart and Information Center, went into great detail as to what Barr and **Greenacre** had witnessed. The department felt the sighting to be of great scientific value.

However, no conclusion was ever made as to the cause of this spectacular light phenomenon.

Base(s)

Bases have long been rumored to be on the Moon, both man-made and **extraterrestrial**. Extraterrestrials are thought by some to have a secret base on the **far side** of the Moon. Earth's governments are also believed by some to have established a base or bases on the Moon. According to one tale, the Germans allegedly established one in the 1940s. Purportedly, there is a base by the name of Luna located on the far side. There is a story of a NASA employee that discovered a base in **Apollo** photographs in the 1960s. Reportedly, the pictures showed an extensive complex that had structures and constructions in a variety of shapes and sizes; including geometric shapes, towers, mushroom- and spherical-shaped buildings. See **Extraterrestrial Bases, Moon Base**.

Bean, Alan LaVern

1932–2018

A former NASA astronaut. He was the Lunar Module pilot on ***Apollo 12*** and served as the commander of *Skylab 3*. A photograph of Bean that was taken during the *Apollo 12* mission has been a source of discussion due to a mysterious image in the picture. The photograph shows astronaut **Charles Conrad Jr**., who was taking pictures at the time, reflected in Bean's helmet visor. Also reflected is a strange **object** with a geometric shape hovering in the air above the lunar floor. Its shadow can be seen in the helmet as well. The **object** appears to be hovering inside an encircling translucent edifice.

In June of 1981, Bean left NASA and took up painting full-time. It was Bean's dream to portray in art what he had witnessed in space and on the Moon. Bean attempted to bring to life the images of what he had seen and present them to the public. He felt that he had experienced something rare and attempted to recreate what he had seen on the missions. Some researchers maintain that the Moon is not simply dull gray, but has more color in certain places than people realize. For clarity, some look to Bean's work. Many of his works of the Moon are filled with color.

Before the Moon

See **Pre-Moon Earth**.

Bellamy, Hans Schindler

1901–1982

Hans Schindler Bellamy was a writer and researcher that authored a number of books based primarily on the studies of Austrian cosmologist Hanns Hoerbiger and German selenographer Philipp Fauth. His books include such notable works as, *Moons, Myths and Man (1936)*; *Built Before the Flood: The Problem of the Tiahuanaco Ruins (1943); The Atlantis Myth (1948); A Life History of Our Earth: Based on the Geological Application of Hoerbiger's Theory (1951);* and *The Calendar of Tiahuanaco: A Disquisition on the Time Measuring System of the Oldest Civilization in the World (1956)*. In 1956, Bellamy introduced the idea that the Moon had come into Earth's orbit approximately 12,000 years ago. He made this conclusion from his studies of the **Sun Gate at Tiahuanaco** in Bolivia. The Sun Gate was filled with symbols containing an astronomical calendar. The calendar disclosed that during that period, Earth's solar year was different from ours today. During that time, per the calendar, the Earth had 290 days in a year, as opposed to the current 365 days. Surprisingly, the information also conveyed that during that era, the Earth had a much smaller moon that was positioned closer to the one we know today. Eventually, the Moon we know arrived, causing chaos on Earth and a reset of the calendar. His findings were published in his book *The Calendar of Tiahuanaco*.

Bergquist, N.O.
(Unknown birth and death dates)

N. O. Bergquist was a Swedish engineer. He is the author of the 1954 book *Moon Puzzle*. Bergquist promoted the idea that the Moon came to be via **catastrophism**. He believed that millions of years ago, the Moon was created when a gigantic **planetoid** hit the Earth in the same direction that the Earth was revolving. The collision caused an expulsion, which propelled a substantial amount of Earth's mass into space. This mass, he deduced, eventually formed the Moon, leaving behind a massive chasm on the Earth which ultimately filled with water. This area today is the Pacific Ocean.

Bessel (Crater)

A small impact crater. It has been described as bright, spherical and measuring 14 miles in diameter. It is situated on the southern half of the **Sea of Serenity** (Mare Serenitatis). On June 17, 1877, English astronomer **Frank Dennet** witnessed a mysterious light hovering over the **crater**. It is the only **transient lunar phenomenon** ever reported near Bessel.

Bible
See **Pre-Lunar Earth**.

Big Ben

An alleged anomalous object on the Moon discovered in a NASA image by researcher Robert Morningstar.

Big Splash Theory
See **Giant Impact Theory**.

Big Whack Theory
See **Giant Impact Theory**.

Binary Planet System

Two celestial bodies orbiting one another whose sizes are so close to each other that it would be out of proportion to refer to one as a moon and the other a planet. In the past, the Moon and the Earth were thought to be a binary planet system, due to the Moon's size in relation to the Earth's. The Moon's diameter is almost 27% that of the Earth's. This is relatively large in comparison to the other planets with moons. Volume-wise, the Moon is about 7.3% of the Earth. The Earth-Moon binary planet system theory was eventually dismissed.

Binder, Otto
1911–1974

A former NASA employee and author. He wrote in the areas of science fiction, science, space travel, **extraterrestrials** and **UFOs**. He was also writer for the Marvel Comics series. There is a mysterious tale that allegedly came from Binder that has made its way into books, around the Internet and into articles. According to Binder, a conversation was recorded by radio hams with VHF receiving facilities. The exchange has become popular among UFO researchers and conspiracy theorists. In his book *Above Top Secret* (page 384), author Timothy Good incorporates the tale. Good writes, "According to former NASA employee Otto Binder, unnamed radio hams receiving facilities that bypassed NASA's broadcasting outlets picked up the following exchange:

Mission Control: What's there? Mission control calling ***Apollo 11***.

Apollo 11: These babies are huge, sir... enormous...Oh, God you wouldn't believe it! I'm telling you there are other spacecraft out here...lined up on the far side the cra**ter** edge... they're on the moon watching..."

Skeptics question the validity of this story, while others believe the incident to be true. See ***Apollo 11*** and **Neil Armstrong**.

Birt (Crater)

According to Don Wilson in *Our Mysterious Spaceship Moon*, in Volume 20 of the *Astronomical Register* (Britain's earliest periodical for amateur astronomers), there was a reported sighting near the Birt Crater of a lunar formation that appeared in photographs to be an image of a sword. The journal states, "Near the crater Birt...is an object shaped like a sword." Charles Fort also commented on the phenomenon in his work *New Lands*, "There is one especial object upon the Moon that has been described and photographed and sketched so often that I shall not go into the subject. It is an object shaped like a sword, near the crater Birt. Anyone with an impression of the transept of a cathedral, may see the architectural here. Or it may be a mound similar to the mounds of North America that have so logically been attributed to the Mound Builders." It was later concluded that the sword shaped object is a lunar fault that lies between the Birt Crater and the Thebit Crater and is more than 75 miles long.

Black Line Phenomenon

There have been reports of a strange reoccurring black line moving vertically across the Moon. People from California's Mojave Desert to the UK have witnessed this phenomenon for years. It has been seen both with telescopes and binoculars, appearing to be present most often during a full or super moon. One witness wrote of seeing the line during a super moon as he drove on a highway in the Northern UK. He stated that it resembled a line drawn with a black marker. Another person described it as moving and flickering. There have been a number of speculations as to what the line could be. These include: an atmospheric phenomenon, an optical illusion and the work of **extraterrestrials**. Some suggest that this may be the same phenomenon that is found in the *Koran*, when the Prophet Mohammed performs the miracle of the "**splitting of the moon**." The most prevalent theory is that the line is from contrails. Some of those that have witnessed the phenomenon do not agree with the contrails hypothesis. They contend that contrails have jagged edges, are extremely volatile, break apart, dissipate and at times change position. The mysterious black line has none of these characteristics, as it remains uniform and stays within the Moon's boundaries.

Black Prince Probe

A mysterious **object** that has been spotted near the Moon on several occasions. It is thought by some to be a probe sent to study and possibly communicate with Earth's occupants. Science fiction writer Alexander Kazantsev postulated that the Black Prince probe may have been sent by extraterrestrials with a message. An American astronomer named Steven Slayton reported on the probe in 1958. He had seen it in the vicinity of the Moon. He was observing the Moon through a telescope when he witnessed a dark, spherical object racing across the sky. The object followed a direct path and eventually vanished after reaching the Moon's perimeter. Slayton contended that the object was anomalous. After Slayton's report, the government allegedly attempted to locate the probe to no avail.

Blair Cuspids

Six formations that were captured in photographs taken by NASA's ***Lunar Orbiter 2*** in the Moon's **Sea of Tranquility**. The cuspids have been compared in appearance to the Washington **monument** and the Egyptian **obelisks**. They are said to have precise shapes and patterns and are in perfectly aligned geometric positions. They are named after anthropologist William Blair who worked for the Boeing Institute of Biotechnology examining the photographs taken by *Lunar Orbiter 2*. His comments about the cuspids appeared in a *Washington Post* article titled "Six Mysterious Statuesque Shadows Photographed on the Moon by Orbiter," published on November 22, 1966. In the article Blair stated, "If the cuspids really were the result of some

geophysical event, it would be natural to expect to see them distributed at random. As a result, the triangulation would be scalene or irregular, whereas those concerning the lunar object lead to a basilary system, with coordinates x,y,z to the right angle, six isosceles triangles and two axes consisting of three points each."

Blair, William
See **Blair Cuspids, Monuments**.

Blood Moon

A blood moon occurs when there is a total **lunar eclipse**. Even though the Earth completely blocks the Sun's rays so they cannot directly reach the Moon during a total lunar eclipse, the rays undergo a scattering effect as they pass through the Earth's atmosphere. This scattering allows some portion of the rays to be deflected towards the Moon. The rays having gone through the Earth's atmosphere also get filtered by the large dust particles that are present close to the surface of the Earth, allowing only the lower frequencies of the Sun's light (reddish colors) to pass through. (This is the same filtering effect that causes the sun to look reddish during sunsets.) Consequently, the filtered, scattered rays of the sun' rays reaching the Moon give the impression that the Moon is crimson-colored. Scientists refer to four blood moons in row a as a "**tetrad**." Even though they are atypical, some centuries occasionally will have a succession of four blood moons.

Blue Flames
See **Blue Lights**.

Blue Light(s)

A form of **transient lunar phenomenon** photographed by ***Apollo 14*** and ***Apollo 16*** astronauts. While in the Fra Mauro Highlands (also known as the Fra Mauro Formation), the **astronauts** took several pictures that showed mysterious images, including blue lights. These lights, also referred to as **blue flames**, took on different shapes and can be found in different points in several photographs. Photographic images of the blue lights taken by the *Apollo 14* astronauts are *AS14-66-9301, AS14-66-9295* and *AS14-66-9299*. A picture taken from the *Apollo 16* mission with the blue lights is *AS16-113-18339*.

Blumrich, Josef Franz
1913–2002

An Austrian space engineer that worked with NASA from 1959-1974. He was also the author of several books. He is best known for his iconic work *The Spaceships of Ezekiel* (1973). After researching and studying the topic **of extraterrestrials** for 18 months, Blumrich determined that they had consistently visited both the Moon and the Earth. In 1974 Blumrich contributed a synopsis of his ideas to the *UNESCO Science Report* (a global monitoring report published by the United Nations Educational, Scientific and Cultural Organization) titled "Impact of Science on

Society." He believed that at some point, humans would soon discover relics left on the Moon by **extraterrestrials**.

Bode, Johann Elert

1747 1826

Prominent German astronomer well known for popularizing the Titius-Bode law and for establishing the orbit of Uranus and naming the planet. In 1774, Bode established the distinguished *Astronomisches Jahrbuch (Astronomic Yearbook),* which he compiled and distributed. Bode studied the Moon and witnessed a number of strange and mysterious lights on the lunar **surface**. He reported on unusual sightings in and around the **Aristarchus crater** from the years 1788 to 1792. On April 9 of 1788, Bode observed that **Aristarchus** appeared "extraordinarily bright" for approximately an hour. From April 9 to 11, of that same year, Bode and fellow astronomer **Johann Schroeter** observed an odd bright spot north the crater rim. From May 8 to 9, also in 1788, Bode monitored strange small brilliantly lit spots of light inside the crater. In March, April and May of 1789 (specific days unknown), Bode saw several illuminated orbs of lights near and around Aristarchus. The mysterious activity that Bode witnessed remains unexplained.

Bogota, Columbia

Indigenous people living in the Bogota, Columbia area remember their ancient history through oral storytelling. One story in particular recounts a period in Earth's history when there was no moon in the sky. A line from one of their legends states, "In the earliest times, when the moon was not yet in the heavens…" The story lends credence to the idea that Earth's night sky was once without a Moon.

Bolivian Symbols

See **Sun Gate at Tiahuanaco**.

Bombing

On October 9, 2009, in a dual mission, NASA sent two spacecraft to the Moon. The first was the probe the ***Lunar Reconnaissance Orbiter*** *(LRO)*. Its job was to map the lunar surface. The *LRO* is said to have been about the size of a car. Next was the *Lunar Crater Observation and Sensing Satellite* (*LCROSS*), which was smaller than the *LRO*. *LCROSS* was sent on a collision course to impact near the Moon's South Pole in an area near the Cabeus crater, located on the Moon's **far side**. There is a conspiracy theory that states that, with this mission, NASA deliberately bombed the Moon to destroy alleged **ruins** from an **extraterrestrial base** that was once located there. Some believe that NASA wanted to bomb them so space programs from other countries would not discover them. NASA's official story is that they were trying to locate water on the Moon. Conspiracy theorists argue that NASA has known about water on the Moon since it was first discovered by **India's** spacecraft ***Chandrayaan-1***. Some believe that *LCROSS* carried a nuclear device and that the ruins were destroyed.

Borman II, Frank Frederick

1928–

A former NASA astronaut. Prior to working at NASA, he was an Air Force officer, aeronautical engineer and a test pilot. Borman worked on the ***Gemini 7*** and ***Apollo 8*** missions. He is best known as the commander of ***Apollo 8***, which made history as the first mission to orbit the Moon. He is also a recipient of the Congressional Space Medal of Honor. In *Our Mysterious Spaceship Moon*, author Don Wilson expressed concern over a mysterious sighting that Borman had during his time on the *Gemini 7*. States Wilson, "there are three visual sightings made by the **astronauts** while in orbit which, in the judgment of the writer, have not been adequately explained." One of those sightings was reported by Frank Borman.

Wilson states, "*Gemini 7* astronaut Borman saw what he referred to as a 'bogey' flying in formation with the spacecraft." What the 'bogey' was remains undetermined. Borman once commented on his experience seeing the Moon while in space stating, "It's a vast, lonely, forbidding expanse of nothing rather like clouds and clouds of pumice stone. And it certainly does not appear to be a very inviting place to live or work."

Born Alongside Earth Theory

A theory that the Moon was created along with the Earth, at the same time, from the same cosmic cloud of gas.

Bottomless Craters

There are **craters** on the Moon whose bottoms have not been located. Some deem them to be bottomless. Lunar scientists suggest that all craters have a floor and contend that due to the darkness and the depth, they are not always easily located. The deepest known crater is five miles down. Some UFO researchers maintain that there are indeed bottomless craters and that these may be access ways to the Moon's interior used by lunar inhabitants.

Boxed Structure

A massive mysterious rectangular structure that was reportedly witnessed in a live broadcast from the Moon when the *Apollo 17* astronauts **Eugene Cernan** and **Harrison Schmitt** were visiting the **Taurus-Littrow** valley, traveling in the *Lunar Rover*. As the story goes, millions of people saw the structure at the time and *CBS News* broadcaster **Walter Cronkite** said live on air that what appeared in the live stream looked like a man-made object. The structure is described as being rectangular, slick and made from what appeared be the same materials that form the lunar soil. The feed of the incident was cut and twenty minutes later Walter Cronkite announced that the camera attached to the *Lunar Rover* had inadvertently photographed itself. There are some who believe that the **astronauts** saw an extraterrestrial structure that day. An analysis of the account can be found at :Alien Moon Structures–Apollo 17, https://www.youtube.com/watch?v=lVyaWOlnuo0.

Bracewell, Ronald N.

1921–2007

Ronald Bracewell was a renowned astrophysicist and Professor Emeritus of Electrical Engineering at Stanford University. In 1960, he published a report on the **Black Prince**, a mysterious **object** that has been spotted near the Moon. Bracewell was of the opinion that this was a probe that was sent from beings from another world that wanted to communicate with the people of Earth, sometime in the future.

Brandenburg, John

(Unknown birth and death dates)

A plasma physicist and defense department expert. He is also the author of *Life and Death on Mars* and other books. Brandenburg was the deputy Manager of the *Clementine* spacecraft mission to the Moon. In referring to the *Clementine* mission, Brandenburg once made mysterious statements in an interview in the SYFY **UFO** documentary *Government Secret Truth Exposed, Aliens on the Moon* (2015). In the documentary, he indicated that the reason the *Clementine* mission went up was to see if someone was creating **bases** on the Moon. He referred to it as a "photo reconnaissance mission." Brandenburg commented about photographs that contained what appeared to be structures on the lunar **surface**, which included what looked to be a large complex, a capsule, large wheels and **domes**. He stated that this indicates that somebody is creating a base. Brandenburg was of the opinion that it could be a potential danger. He continued his analysis of the *Clementine* mission referring to the pictures, stating that he was struck by a picture of "a mile wide recto-linear structure." He believed that it looked unnatural and out of place. States Brandenburg, "I look on any such structure on the moon with great concern because it isn't ours, there's no way we could have built such a thing. It means someone else is up there."

Brass
See **Moon Rocks**.

Breakaway Civilization

A **colony** of people that were once a part of a society, that were recruited to be relocated to a different place, usually by clandestine means. The Moon is believed by some to have a breakaway civilization of humans that either volunteered or were recruited from Earth. There are two theories as to why there would be a breakaway civilization on the Moon. The first iinvolves a worldwide catastrophe. The goal of this alleged civilization is to have people living off of planet Earth in the event of a worldwide situation where the human race could not be saved, be it a natural disaster, a nuclear war or something more nefarious such as an alien invasion. Another reason for there being a breakaway civilization on the Moon is to be able to discern whether or not humans can exist in other places than Earth. In this case, the breakaway civilization would learn what it takes to live away from Earth. The Moon and also Mars are given as places to begin such a practice. The people in these colonies would be studied and monitored to see how they fared outside of Earth in an alien environment. It is believed by some that a breakaway civilization may be operating from inside the Moon's interior at present day. There are a number of ways that people allegedly become a part of a breakaway civilization. Some, it is believed, are chosen before birth in a selection process. Others are asked. Some are chosen due to their expertise in a particular area that would be beneficial to establishing a colony and yet others are thought to be taken.

Bridge

What was believed to have been an artificial bridge on the Moon was discovered on July 29, 1953 by American astronomer **John O'Neill**, the science editor for the *New York Herald Tribune*. The structure was twelve miles long and connected two mountain peaks located on the rim of **Mare Crisium** (Sea of Crises). O'Neill was mocked by his peers at the time. However, British astronomer **Hugh Percy Wilkins** was called upon by O'Neill to examine the area in question and relay his finding. Wilkins reported seeing the structure, confirming the existence of the bridge. English astronomer **Patrick Moore** stated that the bridge had appeared out of nowhere, as this area of the Moon had been studied before and there was no bridge there at the time. The bridge disappeared after a few years. The sudden disappearance of the bridge led to speculation over what happened to it. Some believed that the bridge was dismantled by **extraterrestrials,** possibly to keep their existence a secret from humans on Earth.

Bright Spots

Bright spots on the moon are an anomaly that has been reported by astronomers for centuries. No one can explain what these bright spots are, or where they come from. They have been witnessed by astronomers from around the world and are sometimes seen moving over the lunar **surface**. In an unusual conversation between the **astronauts** of ***Apollo 17*,** there is a mention of bright spots. The Command Module pilot states: "Hey, I can see a bright spot down on the landing site where they might have blown off some of that halo stuff." Some believe that these spots or lights are originated by **extraterrestrials** and the conversation with the astronauts proves to some that extraterrestrials were present as the astronauts went about their work on that final journey to the Moon.

Brighter Moon

There was a time when the Moon shined brighter than it does today. It is said that in the ancient past, the Moon's brilliance was nearly equal to that of the Sun. It not only shined brighter, but it was also said to have appeared considerably larger in the sky. The Catholic Missionary Priest Bernardino de Sahagún, known as the world's first anthropologist, was given the same information about the Moon shining brighter by the natives of New Spain where he studied, worked as an ethnographer and evangelized for over fifty years. An ancient Judaic text supports the idea that the Sun and the Moon once were the same in terms of luminosity. Japan's *Nihongi Chronicles* spoke of a time when the Moon was as bright as the sun, stating "the radiance of the moon was next to that of the sun in splendor." Today the Moon is said to be moving away from the Earth by an inch a year. Therefore, one of the theories for this brighter Moon is that there was once a time when it was closer to Earth than it is today. If it were, then the Moon would have seemed to the people of that period, even larger than the Sun.

British Astronomers
See **William R. Brooks; Cecil Maxwell Cade; Frederick William Herschel; Howard Hill; Walter Goodacre; Nevil Maskelyne; Patrick Moore; Harold T. Wilkins**.

British Selenographers
See **Elger, Thomas Gwyn; Wilkins, Hugh Percy**.

Brooks, William Robert
1844–1921

Noted American astronomer. In 1896, William R. Brooks, who was the director of the Smith Observatory and a professor of astronomy at Hobart College, observed a fast moving circular **UFO** that he described as dark, approximately one-thirtieth of the Moon's length, moving diagonally over the face of the Moon. This mysterious object may have been seen before by a Dutch astronomer named Muller on April 4, 1894, who told of seeing a similar UFO. There was no word as to what this object was or if it was ever seen again.

Bruhl, Hans Moritz von
See **Aristarchus Crater**.

Bruno, Giordano
1548–1600

Italian philosopher and cosmologist theorist. In his work titled *De Immenso* (Bk IV, x, pp. 56-57), Bruno spoke of a time when the Earth had no moon. His comment has been used in the argument for the Moon being an unnatural **satellite**. Bruno writes: "There are those who have believed that there was a certain time (as our Mythologian says) when the moon, which was believed to be younger than the sun, was not yet created. The **Arcadians**, who dwelt not far from the Po, are believed to have been in existence before it [the moon)]."

C

Cade, Cecil Maxwell
1918–1985

British physicist, scientist and astronomer. Once a member of the prestigious Royal Astronomical Society, Cade wrote about seeing strange lights on the Moon. In his book *Other Worlds Than Ours: The Problem of Life in the Universe* (1966), he spoke of anomalous lights and their possible meaning. He states, "Star-like lights, which could not have been due to the Sun's rays illuminating the tops of high mountains, have been the subject of many hundreds of observations; in fact, up to April

1871, no fewer than 1600 observations had been made for the crater Plato alone." In the end, Cade questions what the lights could mean asking, "Were these attempts at signaling by the inhabitants of, or visitors to, the Moon?"

Calendar of Tiahuanaco

See **Hans Schindler Bellamy, Sun Gate at Tiahuanaco**.

Cameron, Winifred Satwell

1918–2016

A former NASA astronomer. The Moon was her passion and her main area of study. Cameron created the largest database of **transient lunar phenomena** ever published. She collected 900 accounts TLP sightings ranging from 1540 to 1970. Her collection of TLPs included everything from unexplained glowing lights to strange shadows, flashes of lights, lights moving across the **surface** of the moon and more. Her work is still used today in research and discussions about TLPs, one of the Moon's greatest mysteries.

Candi

An Indian goddess and the female counterpart to Chandra (ancient Hindu god of the Moon). The two took turns ruling the Moon.

Cape Town Observatory

An astronomer from the Cape Town Observatory once reported seeing a bright light that had appeared in a darkened area of the Moon. He also observed three additional lights in the same region, stating that they were smaller in size.

Capture Theory

The theory that the Moon was captured in Earth's gravity and thus became Earth's **satellite**. Proponents of this theory contend that the Moon was propelled off its original course into an immense elongated trajectory traveling around the Sun. Schindler Bellamy, too, concluded from his interpretation of the symbols on the **Sun Gate** at Tiahuanaco in Bolivia, that the Moon was a planetary object that traveled too closely to the Earth and became caught by its gravitational forces. He believed that this happened approximately twelve thousand years ago. The capture theory hypothesis was eventually dismissed. Some researchers maintain that the Moon is too large to have become caught in Earth's orbit. In comparing the two, in proportion to the Earth, the Moon is around one eightieth of Earth's mass. This is said to make it much too big for it to have become ensnared by Earth's gravity. Popular researcher and sci-fi author **Isaac Asimov** supported this idea. He wrote, "What in blazes is our Moon doing way out there? It's too far out to be a true **satellite** of Earth, it is too big to have been captured by the Earth. The chances of such a capture having been effected and the Moon then having taken up a nearly circular orbit about the Earth are too small to make such an eventuality credible."

Carpenter, Malcom Scott

1925–2013

A former NASA astronaut. He was one of the original seven **astronauts** chosen to serve in NASA's Project Mercury (the first human spaceflight program and a prelude to the **Apollo missions** to the Moon) in 1959. He has the distinction of being the second American to orbit the Earth. He was also the fourth American to work in space. Carpenter once made this cryptic comment, "At no time when the astronauts were in space were they alone: there was a constant surveillance by **UFOs**." It has been speculated that there is a message behind Carpenter's words. Some wonder if he had seen more in space than he told and if he took secrets concerning the Moon with him when he passed away.

Cartographer(s)
See **Edward Barr; James Greenacre**.

Cassini (Crater)
See **Gian Domenico Cassini**.

Cassini, Gian Domenico
(Also Jean Dominique Cassini)
1625–1712

A noted Italian (born French) astronomer, engineer and mathematician. He was once the director of the Paris Observatory. He also charted the moon. Cassini is connected to several scientific breakthroughs and ventures, including the first research and analysis of Saturn's moons. The Moon's Cassini **crater** is named in his honor, as well as the Cassini spacecraft. In 1671 Cassini experienced an anomalous event on the Moon that remains a mystery. As he lifted his telescope that day, he was perplexed to see what he later described as a white cloud floating above the **surface** of the moon. He offered no explanation as to what he thought it might be, but was careful to record the anomaly.

Castle

A strange anomaly located in ***Apollo 10*** photographs. The **object** is suspended approximately seven miles above the lunar **surface** and has the appearance of a ruin of an ancient castle. It has been named "the castle" by researcher and writer Richard C. Hoagland. The bottom of the object seems to have a row of pillars and what looks to be a turret on the top. The NASA picture in which the castle can be seen is AS10-32-4822.

Cavern(s)
See **Lunar Cavern(s).**

Censorinus
(Unknown birth and death dates)

A third century Roman grammarian and author. In his writing titled *De Die Natali*, Censorinus writes about a time in Earth's history when there was no Moon in the sky. His writing has been used by some Moon researchers to support the hypothesis that the Moon is not a natural **satellite**, and adds to the mystery of the Moon's origin.

Censoring of Information Conspiracy Theory
The theory that there is a cover-up regarding what the **astronauts** saw on the Moon.

Cernan, Eugene
1934–

Former NASA **astronaut**. Cernan also worked as an aeronautical engineer and a fighter pilot. He was the last man to walk on the Moon. His journeys into space included three missions. In June of 1966, he served on ***Gemini 9A*** as co-pilot. In May of 1969, he was ***Apollo 10s*** Lunar Module pilot. In December 1972, on his final mission, he was the commander of ***Apollo 17*** (the last Apollo Moon mission). As they landed on the Moon, Cernan exclaimed, "We is here, man! We is here!" On that assignment he and astronaut **Harrison Schmitt** studied the **Taurus-Littrow** area. There are strange tales of the *Apollo 17* **astronauts** seeing **extraterrestrials** and extraterrestrial structures while on the Moon's **surface**. **UFO** enthusiasts claim that certain quotes from Cernan allude to him witnessing something out of the ordinary in space. They cite his speech from the Moon on his last day when he stated, "I'm on the surface; and, as I take man's last step from the surface, back home for some time to come—but we believe not too long into the future—I'd like to just [say] what I believe history will record. That America's challenge of today has forged man's destiny of tomorrow. And, as we leave the Moon at Taurus-Littrow, we leave as we came and, God willing, as we shall return: with peace and hope for all mankind. Godspeed the crew of *Apollo 17*." Additionally, Cernan was once quoted as saying, "maybe the moon can tell us something about the existence of some ancient civilization, not necessarily on earth, nor necessarily on the moon, but possibly within our own universe and give us insight into what reality is all about." See ***Apollo 17;*** **Boxed Structure**.

Chalcidensis, Dionysius
See **Pre-Lunar Earth**.

Chaldea(ns)

Chaldea was an ancient province of Babylonia. In ancient Chaldean lore, there is a legendary tale about the **origin** of the Moon. The tale seems to offer an explanation as to how the Moon came to be Earth's **satellite**. It states, "Anu opened up the original abyss and created a whirling motion like boiling. The Moon passed through an opening like a giant bubble and made its way across the heavens." To some, that passage appears to suggest that the Moon was mysteriously brought into our solar system from elsewhere. Others wonder if the "opening" was a kind of gigantic portal where the Moon came through, or was sent through into our universe.

Chalet, The

An anomalous construction on the **Tycho crater**. An image of the item is featured in author **Mike Bara's** book *Ancient Aliens on the Moon* (page 186). In his book Bara states that the object "looks somewhat like an A-frame building." Bara located the "Chalet" in a photograph taken by the ***Clementine*** spacecraft.

Chandrayaan-1

India's first lunar probe. *Chandrayaan-1* was launched on October 22, 2008 by the Indian Space Research Organisation (ISRO). It was expected to perform for two years. However, in August of 2009, *Chandrayaan-1* suddenly lost contact in the middle of transmitting data back to Earth. At that point, *Chandrayann-1* had completed ninety-five percent of its mission. Unable to reestablish contact with Chandrayann-1, the ISRO announced that *Chandrayaan-1's* mission had come to an end. In 2017, NASA located *Chandrayaan-1* still orbiting the Moon. It was *Chandrayaan-1* that first located **water** on the Moon. Strangely, *Chandrayann-1* also located signatures of organic matter (material that was a part of a recently living organism) on the lunar **surface**. Just where this mysterious organic matter originated and what it is from remains a mystery.

Chang-E
(Also Chang-O)

Chinese moon goddess that resides on the Moon. Several Chinese spacecrafts were named in her honor. See **China**.

Chang'e-2
See **China**.

Chang'e-3
See **China**.

Charroux, Robert
1909–1978

A popular French historian and author. He was best known for being a proponent of the ancient **astronaut** theory. His works include *One Hundred Thousand Years of Man's Unknown History* (1963), *Forgotten Worlds* (1973), *Masters of the World* (1974), *The Gods Unknown* (1974) and *Legacy of the Gods* (1974). In his book *One Hundred Thousand Years of Man's Unknown History,* Charroux states, "Are we to conclude that these lunar **craters** have been frequented by extraterrestrial astronauts?... The possibility cannot be rejected, especially with regard to the **Plato crater**, where many mysterious lights have been observed."

Charroux's name is associated with a mysterious moon incident. **Apollo** astronaut **Alfred Worden** received an eerie radio message from the Moon that he forwarded to NASA. The message is said to have been censored and halted from being released. The language was in an unknown dialect and not related to anything on Earth. Oddly, the message was later broadcast in the media in France. Charroux was suspected to have been the person that leaked the message. Professional linguists investigated the transmission, but were unable to interpret it.

Cheshire Moon
See **Wet Moon**.

Childress, David Hatcher
1957–

American researcher, author, lecturer and publisher. He is the owner of Adventures Unlimited Press, which specializes in books on such esoteric topics as ancient mysteries, alternative history and unexplained phenomena. Childress has written about such topics as **UFO's** and ancient astronauts. His book *Extraterrestrial Archaeology* includes the latest discoveries about the Moon. Says Childress, "Lights, clouds, apparent structures and mysterious streaks have fueled speculation on the mysteries of the Moon for hundreds of years."

China

Early studies of the Moon and lunar phenomena in China began millenia ago. Ancient Chinese cosmologists recoded the Moon's movements, which contributed to their formation of the Chinese zodiac. In addition, ancient Chinese research made it possible to predict **lunar eclipses**. China has also had its share of mysterious phenomena involving the Moon. United States astronaut **James McDivitt** once spotted a **UFO** as he flew over China. McDivitt said that the UFO was moving low over China and described it as a "bright star, moving fast."

China's lunar orbiting spacecraft, *Chang'e-1* was launched on October 24, 2007. The success of *Chang'e-1* paved the way for future Chang'e missions. China's first spacecraft to land on the Moon was *Chang'e-3*. It was launched in December 2013. During the mission, *Chang'e-3* filmed a UFO that streamed an array of colors, illuminating the Moon's **surface** as it passed by. According to the pictures the UFO was enormous. Sources have it that it took a little over six minutes to cross over. The pictures showed a great deal about the object as it passed over *Chang'e-3*. They revealed how the UFO looked, the moment that it appeared and its movement as it crossed over the Moon. There are also tales of China's *Chang'e-2* photographing what is speculated to be an outpost with an unnatural environment on the Moon. Rumors have it that *Chang'e-2* took pictures of buildings that had been partially destroyed. In 2018 it was announced that China was constructing a manned spacecraft that could send up to six **astronauts** to the Moon.

In the book *Invader Moon* (page 166), author Rob Shelsky writes about some of the artificial **objects** the Chinese allegedly found when orbiting the Moon. Shelsky writes, "After orbiting a satellite around the Moon, the Chinese released some interesting photos. These show objects that do not appear to be natural. In other words, they appear artificial. Moreover, they are not the product of humans."

Chronological Catalog of Reported Lunar Events

An official lunar anomalies report from NASA detailing **transient lunar phenomena or TLPs** (also known as lunar transient phenomena or LTPs). TLP's are irregular, short-lived flashes of light and unexplained illuminations on the Moon. Hundreds have been reported by **astronomers**, scientists and **astronauts**. The report was commissioned in the 1960s by NASA and released to the public in 1968. It references four centuries worth of sightings recorded by over 300 people, of strange, anomalous and mysterious phenomena observed on the Moon. The catalog includes anomalous lights, transitory changes, unusual colors on the lunar floor, the appearance of glowing mists, flashes of light, potential volcanic movement and **craters** resembling **domes**. Each recorded observation contains the date, description and name of the astronomer or observer, along with a reference. The catalogue includes over 570 temporary abnormalities. States the report, "the purpose of this catalog is to provide a listing of historic and modern records that may be useful in investigations of possible activity on the moon." NASA wanted a report that would chronologically list Moon anomalies that were reported by astronomers from 1540 to 1967. It was composed and written by Barbara Middlehurst, of the University of Arizona; Jaylee M. Burley, of the Goddard Space Flight Center; **Patrick Moore**, of the Armagh Planetarium; and Barbara L. Welther, of the Smithsonian Astrophysical Observatory. The report can also be found under the names *Document R 277* and *NASA Technical Report R-277*.

Cigar-Shaped UFOs
See **UFOs**.

Circumference

The Moon's circumference at its equator is 6,790 miles.

City Ship

An alleged, gigantic **extraterrestrial** spacecraft designed to hold thousands of people at a time, housing them as they travel the universe. Its purpose is to transport people between solar systems and galaxies. The people on board can stay for long periods of time, because the ship acts as a temporary home, complete with everything needed for supporting life. There is a story of an individual capturing a picture of a large ship hovering over the Moon. The person who took the photograph was said to be an experienced **UFO** investigator and a sergeant for the US Army. He believed that the huge, silver colored, spherical spacecraft that he had captured on film was no less than a city ship.

Clementine Spacecraft

A spacecraft sent to the Moon on a combined mission between the Ballistic Missile Defense Organization (BMDO) and the National Aeronautics and Space Administration (NASA). It launched on January 25, 1994. The name *Clementine* was inspired from the American Western folk ballad *Oh My Darling Clementine*, due to the inevitability of its eventually being lost forever in space once the mission was completed. The line from the song is, "You are lost and gone forever, Dreadful sorry, Clementine." In his book *Ancient Aliens on the Moon*, **Mike Bara** offers an in-depth look at anomalous features found at the **Tycho crater** by *Clementine*, as well as a photograph taken of a mysterious **transient lunar phenomenon**. See **John Brandenburg**; **Moon Base**.

Coaccretion Theory

A premise of how the Moon was espoused by French physicist and astronomer **Pierre-Simon Laplace,** who became famous for his analysis of the stability of the solar system. It is also known as a "**common birth theory**" and "condensation hypothesis." The thought behind the theory is that the Earth and the Moon were created at the same time, yet independently of each other, from a nebular cloud of dust and gas that coalesced over time. The coaccretion theory as being the **origin** of the Moon has been dismissed by scientists.

Coke Bottle

An Australian woman by the name of Una Ronald claims to have seen a Coke bottle when watching the ***Apollo 11*** Moon landing. Supposedly, she saw the Coke bottle kicked across the Moon. The story caused a controversy, some claiming that it was evidence that the

Apollo 11 Moon landing was a hoax.

Collier, Alex

An extraterrestrial **contactee** for the Andromedans, a race of beings from the Andromeda galaxy. His primary contacts were two Andromedans that he knew as Morenae and Vissacus. The extraterrestrials purportedly taught Collier a great deal about galactic history, including that of the Moon. He was given a vast amount of information that he has shared through his writings and lectures. He covers such topics as **extraterrestrials** (including existing races on Earth, advanced beings and the Reptilians), the dark forces, Mars, the new world order and more. During an interview in1994, Collier discussed a number of Moon related issues. He claimed that extraterrestrials brought the Moon into Earth's orbit and that the entities that performed this task exist among us. Collier suggests that there are areas of the Moon that thrive with vegetation. He also believes that the Moon has water as well as an **atmosphere**. Collier once commented, "Our Moon has an atmosphere that is in many respects similar to that of the Earth. In many large craters on the visible and the invisible side, the atmosphere is denser than sea level on Earth, it is claimed."

Collins, Michael

1930–

A former **astronaut** that flew on ***Gemini 10*** as a Pilot and the history making ***Apollo 11***, where he served as the Command Module pilot. He is also a retired Major General for the United States Air Force and formerly was a test pilot. During the *Gemini 10* mission, the **astronauts** reported that two "bright objects" were following their spacecraft. Eventually, the objects zoomed out of sight. Later, the astronauts reported strange objects following them a second time. The astronauts' best guess as to what they could be were satellites. *Apollo 11* brought a whole new set of perplexing outer space experiences to Collins. In one case, as they headed for the Moon, the crew encountered a **UFO** that was trailing them. Reportedly, the UFO was illuminated and was shaped like the letter L. Attempting to understand what they were witnessing, the **astronauts** contacted mission control inquiring about the whereabouts of the *Saturn V* launch rocket. They were informed that the *Saturn V* was around six thousand miles away from their position. Researchers speculate that what the astronauts saw that day was a part of the *Saturn V* or some other piece of equipment. Others maintain that it was quite possibly an **extraterrestrial** spacecraft. Collins once commented on his impression of the Moon, "When the Sun is shining on the **surface** at a very shallow angle, the **craters** cast long shadows and the Moon's surface seems very inhospitable. Forbidding, almost. I did not sense any great invitation on the part of the Moon for us to come into its domain. I sensed more that it was almost a hostile place, a scary place." See **Appendix 1: NASA Transcripts (*Apollo 11, Gemini 10*); Moltke Crater; Plaque**.

Colony

See **Alaje of the Pleiades; Colony on the Moon Conspiracy Theory; Humans on the Moon**.

Colony on the Moon Conspiracy Theory

The theory that people from Earth are secretly living in a colony on the Moon, placed there by one or more of the Earth's governments.

Color

There is a theory that certain areas of the Moon have color and that it is not simply a dull rock. States the popular website *The Greater Picture: The History of Earth*, "The moon is a colorful world." It goes on to say, "Photos of the moon that are publicized by NASA are consistently tampered with to remove any trace of color or life." Conspiracy theorists believe that this is an attempt to keep the public in the dark about the true nature of the Moon and that it is not simply a bare, colorless rock. Former astronaut **Alan Bean** presented his paintings of the Moon to the public. His work shows a vibrant, colorful world. Art was Bean's way of showing everyone what he had seen on the Moon. Rosemary Ellen Guiley states in her book *Moonscapes* (page 182), "Observers on earth and other **Apollo** astronauts have reported a variety of colors on the lunar **surface**. In 1971, ***Apollo 15*** astronaut **James B. Irwin** reported finding **rocks** that were white, black, green and 'all conceivable colors.'" See **Franz von Paula Gruithuisen; Orange Soil.**

Colossus

In 2014 a figure nearly 150 feet tall was spotted in a NASA photograph on Google Moon. The photograph is believed to have been taken by *Apollo 15*. The image went viral when posted to the Internet. Some believe it to be an **extraterrestrial** walking on the lunar **surface**. To some the figure resembles the ancient Greek Colossus of Rhodes; it was therefore dubbed "Colossus." Some question whether it might be an ancient ruin of a statue or monument. There is speculation that alleged ruins found on the Moon are from an ancient civilization that was destroyed in a war. There are those who believe that this figure may have been from that time. The official word is that it is from dirt on the camera lens.

Comet(s)

There are some moon researchers that have the notion that the Moon is not a natural **satellite**. In attempting to explain how the Moon came to be in Earth's orbit if it is unnatural, it has been proposed that it may have been brought in by a comet. Contactee **Alex Collier**, who claims to have received information from advanced **extraterrestrials,** in fact states in his teachings that the Moon was placed in the tail of a comet and dragged into our solar system. Another **contactee** by the name of Billy Meier relays the same tale. According to Meier, the Moon originated as a small planet and was brought into our solar system by a comet.

Common Birth Theory

See **Coaccretion theory**.

Condorcet Hotel

An anomalous construction found on the Moon by the ***Apollo 17*** astronauts. The Condorcet Hotel was mentioned in a conversation between the *Apollo 17* **astronauts** while on the lunar **surface**. The astronauts' reportedly strange conversation added to the mystery of the *Apollo 17* mission, as they spoke of something special they were witnessing. See **Appendix 1: NASA Transcripts (*Apollo 17*); Condorcet Hotel**.

Conrad Jr., Charles "Pete"

1930–1999

A former NASA **astronaut**, aeronautical engineer and naval officer. He was a part of several space missions including ***Apollo 12***, *Gemini 5, Gemini 11* and *Skylab 2*. He was the third man to walk on the Moon, which occurred during the ***Apollo*** *12* mission. As he stepped onto the Moon, Conrad exclaimed, "Whoopie! Man, that may have been a small step for Neil, but that's a long one for me." In a photograph taken by Conrad of astronaut **Alan Bean**, Conrad can be seen in the reflection in the screen of Bean's helmet. A strange **object** can also be seen in the reflection of the helmet's screen. The item is in a geometric pattern and can be seen hovering over the lunar floor. The shadow of the mysterious object can also be seen. **UFO** enthusiasts contend that the shadow in the picture proves that the object was real and that there was not a problem with the photographic equipment. There is no explanation as to what this was and it remains a mystery.

Conspiracy Theory/Conspiracy Theories

The belief that there is a secret plan by an organization to cover up information on a particular topic or event. Conspiracy theories involving the Moon have been around since the space race. Some of the more popular conspiracy theories include whether or not the "Apollo Moon landings" were faked and the question of why there are no longer missions to the Moon. See ***Apollo 11***; ***Apollo 18–20***; **Apollo Program; Neil Armstrong; Otto Binder; Bombing; Censoring of Information Conspiracy Theory; Colony on the Moon Conspiracy Theory; Color; Domes; Endymion Crater; Far Side; Flag Conspiracy Theory; Hologram Conspiracy Theory; Moon Landings Conspiracy Theory; Moon Missions to Locate Ruins Conspiracy Theory; Ovid; Reptilians; Ruins; Warned off the Moon**.

Contactee(s)

Someone that is in communication with **extraterrestrials**. Such a person is generally in contact with the extraterrestrials through telepathy, channeling and in some cases, direct visitation. The main reasons extraterrestrials give for approaching a contactee is that they want to assist mankind and that they want to prevent humans from using nuclear warheads. The extraterrestrials have stated to some that they are here to monitor what is being done in regard to nuclear war, to prevent mankind from damaging the Earth and also the Moon. Some contactees have reported boarding a spaceship and traveling with extraterrestrials to a remote place or another world, and in some cases the Moon. Contactees that claim to have visited the Moon include **George Adamski**, **Howard Menger** and **Buck Nelson**. Contactee **Alex Collier** was given information from extraterrestrials about the Moon.

Cooper Jr., L. Gordon

1927–2004

A NASA **astronaut** that served on the ***Mercury 9*** and *Gemini five* space operations. NASA's Project Mercury and **Project Gemini** engaged in missions that were in preparation for someday putting **astronauts** on the Moon. Cooper began his career as an aerospace engineer, a test Pilot and an Air Force pilot. He was one of the seven original astronauts assigned to Project Mercury. During his professional career, Cooper had mysterious experiences as a civilian and when serving aboard each of the NASA missions. According to reports, in 1963, while in orbit over Hawaii aboard *Mercury 9*, Cooper heard a mysterious voice speaking in an indecipherable language over the radio. This was said to have been on a special radio frequency. The transmission was later studied by analysts and it was determined that the language was not identifiable on Earth. On the same mission, Gordon claimed to have seen a large green **UFO** approaching his craft before suddenly vanishing.

Copernicus (Crater)

A magnificent impact crater that is less than a billion years old. It is named after the brilliant Polish mathematician and astronomer Nicolas Copernicus. Copernicus (1473–1543) is the originator of the hypothesis that the Moon rotates around the Earth and the planets orbit the Sun. The Copernicus crater boasts a dramatic, sprawling ray system that beams out over the surrounding darkened maria. Found in the eastern **Oceanus Procellarum**, Copernicus spans nearly sixty miles across and is two miles deep. Its terrain is full of crags, cliffs, foothills, peaks and valleys. Its mountains range in size from 1500 to 2000 feet high and 10 miles long. Sitting above the rim the crater is a translucent **dome** that radiates a strange bluish-white light emanating from the dome's interior. There have been debates over what may lie beneath the dome and the source of the light. Some **UFO** proponents theorize that there may be a fusion reactor beneath the dome, put there by extraterrestrial beings and that may be the cause of the mysterious light. Additionally, near Copernicus, there is a rectangular area of about 1000 ft by 1200 ft that appears to be unnatural. On November 5, 1954, a strange, unexplained glowing point of light was seen within the crater.

Coppens, Philip
See **Johann Hieronymus Schroeter**.

Cori, Patricia
(1952–)

Spiritual leader, author and public speaker. In her book *Cosmos of the Soul: A Wake-up Call for Humanity,* Cori writes extensively about the Moon. In chapter eleven, titled The Secret Government and the Space Conspiracy, Cori writes about the Moon's mysterious and secretive ancient past.

Cosmonaut(s)

A Soviet astronaut. In the space race between the United States and the Soviet Union, placing a man on the Moon was an objective of both countries. During the 1960s as the Soviet Union prepared to explore the Moon, there were mysterious and sometimes incomprehensible events that occurred. One such story is told in the article "The Moon is a Foreign Nation," by author Steve Omar. According to the story, the Soviets sought to break the record for time in orbit. When the Soviet spacecraft arrived in space, the cosmonauts were followed by **UFOs**, which proceeded to encircle the craft and then in what appears to be a means of preventing them from going any further, began to push them around between them, as if they were in a pinball game. Shocked, the Soviets are said to have abandoned the mission and returned to Earth. Renowned radio broadcaster and author **Frank Edwards**' investigation into Soviet cosmonauts made headlines when he revealed, in June of 1962, that several cosmonauts had lost their lives in space-related incidents. One of the cosmonauts was a woman. In his book *Strange World,* Edwards wrote an article centered on true events that he titled "First Woman in Space." According to the account, On February 17, 1961, a large Soviet booster launched from Baikonour Cosmodrome, a Russian spaceport. According to Edwards, it was believed at the time that the Soviets may have been taking aim at a manned orbit of the Moon. However, the *Luna* capsule never reached enough velocity to break away from the Earth's gravity. Tracking stations of that period picked up and taped the conversations of the two cosmonauts, one male and one female, that were in the capsule. The pair were in space for seven days and could be heard transmitting updates to the Soviet Union. The cosmonauts were saying regularly that all was fine, even though it was apparent that it was not. The scenario came to an end on February 24, 1961. Right after the cosmonauts radioed in

that all was well, things took a turn for the worse. In what would be the last transmission from the cosmonauts, as recorded by tracking stations at Uppsala, Bochum, Turin and Meudon, one could hear a frightening conversation where the male cosmonaut was commenting about not being able to understand signals and how they couldn't view anything. Afterward, the female cosmonaut could be heard anxiously using phrases such as "maintaining equilibrium" and holding on to something "tight." She was yelling to her partner to "look out of the peephole!" The male voice then yelled back that if they didn't get out, "the world will never hear about it." Just after that final statement, things went silent. No word was ever heard from them again. In his book *Strange* World (page 388), Edwards, states, "What happened, will probably never be known, for it is highly probable that the Soviets themselves do not know. We can only hope that whatever fate befell them was merciful and swift."

Cramer, Randy

Randy Cramer tells a strange story of being a part of a program titled the US Secret Space Program (SSP), where he served as a Marine for 20 years. He allegedly served on both the Moon and Mars in fully-equipped installations. He was on Mars for 17 years. After a conflict on Mars, he returned to the Moon where he worked as a military pilot for three years, before returning to Earth. According to Cramer, he has been given permission to speak about his experiences publically.

Crater(s)

The Moon is pockmarked with numerous craters, more than 500,000 of which can be viewed from the Earth. In diameter, they vary significantly, the largest crater being Bailly at 184 miles and the smallest Hipparchus at 93 miles. There are two main theories to explain why the Moon's **surface** is filled with so many craters as compared to the Earth's. The first is that they were created by volcanic activity; the second involves meteorites impacting the surface. Most researchers favor the meteorite supposition, supported by the idea that the Moon has no atmosphere to protect it against meteorites, which have bombarded the lunar surface for billions of years. The collision of meteorites on the surface of the Moon can release immense amounts of energy, some comparable with nuclear explosions. A mystery arises upon examination of the Moon's **craters**. Even though they vary in diameter, they do not vary much in-depth. No matter how large or how fast, the meteorites do not seem to penetrate the **surface** more than 1.2-2 miles, even though they may leave behind a 40-mile wide crater. Scientists are unable to explain this unusual occurrence. See **George Adamski; Alphonsus Crater; Archimedes Crater; Aristarchus Crater; Aristotle Crater; Asada Crater; Bessel Crater; Birt Crater; Bottomless Craters; Cassini Crater; Copernicus Crater; Daedalus Crater; Endymion Crater; Eratosthenes Crater; Eudoxus Crater; Gassendi Crater; Greaves Crater; Grimaldi Crater; Humboldt Crater; Kepler Crater; Linne Crater; Littrow Crater; Lobachevsky Crater; Moltke Crater; Piccolomini Crater; Plato Crater; Pluto Crater; Proclus Crater; Tyco Crater; Waterman Crater**.

Crescent Moon

The image of the Moon as it is seen in its first quarter, which is a bowl shape with edges ending in points. These points are sometimes referred to as "horns of the Moon." During this period, it is said that the Moon appears to be "smiling." In ancient times, lunar gods and goddesses were depicted with crescent moons near them and often on their heads as a crown. The image of the mysterious crescent is beloved around the world and can be found on a number of flags, including those of Algeria, Cyprus

(North), Libya, Malaysia, The Maldives, Nepal, Pakistan, Singapore, South Carolina, Tunisia and Turkey.

There have been a number of unexplained lights seen and reported during the crescent moon over the centuries. As far back as 1587, an astronomer in England reported seeing a bright light between the "horns of the Moon." In November of 1668, Minister **Cotton Mather** wrote a letter to a Mr. Waller of the Royal Society to inform him of his unusual sighting of a "star" between the horns of the Moon.

Cronkite, Walter Leland

1916-2009

Iconic American broadcaster that worked for the *CBS Evening News*. Cronkite witnessed a mysterious object on the Moon during a live broadcast of the *Apollo 17* moon mission. As astronauts **Eugene Cernan** and **Harrison Schmitt** traveled through the **Taurus-Littrow** valley, they allegedly happened across a gigantic box-shaped structure. When witnessing the image, Cronkite reportedly stated that it looked man-made. The feed was then cut. Cronkite later returned to inform the world that the *Lunar Rover* (where the camera was located) had inadvertently taken a picture of itself. It is said that Cronkite was amused by the incident. See ***Apollo 17;*** **Boxed Structure**.

Cross(es)

Structures in the shape of crosses have reportedly been discovered on the Moon. They are said to be "perfectly shaped" and are found outside the rims of many **craters**. Some wonder if the odd constructions are flukes of nature, or if they have a higher meaning, placed there by advanced intelligent beings long ago. Others maintain that the objects are optical illusions, appearing in a cross-like shape due to the combinations of lighting and shadows. In the book *Elder Gods of Antiquity* (page 102), author M. Don Schorn, writes, "author Peter Kolosimo reported that photographs of strange structures, in the shape of a cross, were taken by astronomer R.E. **Curtis** and published in the *Harvard University Review*. They were deemed to be too purposefully arranged to be merely natural formations."

Crust

The outer layer of the Moon's crust is 36 to 62 miles thick.

Crystallized Moon

A number of astronomers have reportedly witnessed seeing stars twinkling in the first lunar phase of a new moon. These starlike lights were seen in the darkened portion of the Moon, when the Moon appears to be just a sliver of light. This odd effect has led some to speculate that the Moon may well be crystallized.

Curtis, Robert E.

(Unknown birth and death dates)

On the evening of November 26, 1956, an astronomer by the name of Robert E. Curtis from Alamogordo, New Mexico inadvertently took a picture of something highly unusual on the Moon. Curtis was studying the Moon, all the while taking pictures. Later, after developing the film, he was stunned to see what appeared to be a large, glowing, perfect white **cross**. Each branch of the cross spanned what he estimated to be several miles. Each point of it was centered at the exact location to form a cross. Scientists tried to explain it away as the edges of mountains crisscrossing at the precise angles to give the illusion of a cross. Author Frank Edwards commented on the subject in his book *Strange World* (page 275) saying, "Unfortunately for that explanation, it is physically impossible for mountain ridges to cross each other at right angles."

Cuspids

See **Blair Cuspids, Monuments**.

Cylindrical-Shaped UFOs

Cylindrical-shaped **UFO**s have been seen in space and around the Moon for years. They have also been spotted on Earth in large numbers. There is one photograph taken from ***Apollo 15*** that shows what appears to be a cylindrical-shaped craft on the Moon. It was this picture that is believed to have inspired the ***Apollo 20* hoax.** See ***Gemini 4***.

D

Daedalus (Crater)

A conspicuous lunar **crater** named after Daedalus of Greek mythology. It dates to the Early Imbrium period (3.75 to 3.2 billion years ago). It is positioned in the area near the center of the far side of the Moon. It is a site where there appears to be a cluster of **domes**. Some believe that these domes house a base underneath.

Dark Side

The area of the Moon that cannot be seen from Earth. It also known as the **far side**. The term "dark side of the Moon" is a misnomer as sunlight hits both sides of the Moon. Author and researcher **Mike Bara** states in his book *Ancient Aliens on the Moon*, "Contrary to popular belief, there is no 'dark side' of the Moon. Because of its constant motion in orbit around the Earth and its own constant twenty-seven-day synchronized spin, at some point in the month the entire lunar **surface** is exposed to the light."

Darwin, George

See **Fission Theory**.

Day

See **Length of a Day**.

Dean, Robert "Bob" Orel

1929–2018

America UFOlogist. Dean served in the US Army as a command sergeant major. He gave presentations on such topics as **extraterrestrial**s, government cover-ups and **UFOs**. At a seminar in Europe one year, Dean vented his frustration with NASA over the alleged destruction of forty rolls of film from the **Apollo program**. The film is said to have contained images of the **astronauts** circling the Moon, **Apollo** landings on the Moon and the astronauts working on the lunar **surface**. Dean maintained that they were destroyed because it was decided that the photographs would be socially and politically disturbing. The mystery of what exactly was on those rolls of film and why they would have been destroyed and why they would be destroyed, remains.

Death Star

In the *Star Wars* movie franchise, the *Death Star* is an artificial moon, space station and weapon. It has the capability of obliterating a planet. The Earth's Moon has been compared to the *Death Star* model. Some ufologists believe that the Moon is artificial and that there is a possibility that it is a space station housing beings that are technologically more advanced and are observing the activities of the people of the Earth. One official has even suggested that there may be a military base on the Moon that could be a potential threat to the Earth.

Delporte (Crater)

See **Apollo 20 Hoax**.

Democritus

460-370 BCE

A Greek philosopher from Abdera, Greece. Democritus taught his pupils about a race of people that lived on Earth before there was a Moon. He referred to them as the **Arcadians**.

Dennet, Frank
(Unknown birth and death dates)

English astronomer who in June of 1877 reportedly saw a mysterious point of light over the **Bessel crater**.

Dennett, Preston
1965–

A **UFO** and paranormal investigator, researcher and author of several books and articles. He is also a field investigator for the Mutual UFO Network (MUFON). In his book *Out of Body Exploring,* Dennett recounts having an out-of-body experience and traveling to the Moon. In the passage, he talks of being surrounded by stars just before seeing the Moon appear before him. He observed the landscape writing that he saw, "rocky fields" and "**craters**." He was surprised to see an enormous edifice that resembled a "dome-covered stadium." He describes the peculiar structure as being camouflaged, as it was closely matching the Moon in color and resembled a hill. Inside the structure he found various kinds of "high-tech equipment." He recalled that the walls were "lined with display screens." On a table sat a cage with a strange being inside that he believes was a part of a scientific experiment. He thought it to be a human child that had been genetically altered.

Density

The Moon's density is 3.34 g/cm^3 (grams per cubic centimeters).

Descartes (Crater)
See ***Apollo 16***.

Diameter

The Moon's diameter is 2,160 miles (approximately 27% of the Earth's diameter of 7,910 miles).

Disappearing

The Moon is progressing very slowly away from the Earth. It is moving away by approximately one inch every year. At this rate, it will be millions of years before the Moon is substantially farther away from the Earth

Disclosure Project

A project founded by ufologist **Steven Greer** seeking full disclosure of information on **extraterrestrials** and **unidentified flying objects** that has been covered up. Researchers maintain that if there is disclosure some of the many mysteries surrounding the Moon may be answered.

Distance from Earth

The mean distance from the Earth to the Moon is 238,857 miles (384,400 km). It took the Apollo missions approximately three days to reach the Moon.

Dome(s)

Dome-like structures, often with lights, have been seen on the Moon by astronomers for centuries. German astronomer **Johann Schroeter,** who had observed domes on the Moon, believed they were the work of Selenites. In modern times, there have been two hundred reports of domes being spotted by Moon observers in a variety of sizes. At one point 20 to 30 domes were seen on the bottom of the **Tycho crater**. They appear and later simply vanish. This has led some researchers to speculate that there is something or someone manufacturing the domes. They believe the domes are coming from a source with advanced power and technology. There are also unexplainable, stationary domes. The Apollo **astronauts** witnessed them and commented on them. These domes generally are seen with glowing lights beneath or inside them, leading some to speculate that there is some activity going on beneath the **surface**. Some maintain that there is a power source under the

An artist's impression of a lunar dome.

lunar floor. In the early 1900s, astronomers in France publically stated that they had observed domes on the Moon. Years later in 1966, NASA released thirty-three Moon dome photographs taken by ***Lunar Orbiter 2***. NASA's *Ranger 2* which was launched on November 18, 1961, reportedly took more than two hundred photographs of **craters** on the Moon with domes inside them. **China's** *Chang'e-3* probe found and photographed what is believed by some to be ancient, shattered lunar domes. The mysterious domes have led conspiracy theorists to believe that there is a cover-up regarding a civilization once existing on the Moon. In addition, some believe that humans have been existing there under a secret space program for years and speculate the domes are evidence of an artificial environment on the surface of the Moon. Others embrace the **hollow moon theory** in which extraterrestrials live beneath the surface in a city. Others maintain that the domes are attached to the **spaceship Moon theory** where again, beings are believed to live inside the Moon, operating the ship and observing humans in the process. Some **UFO** enthusiasts and conspiracy theorists insist that several domes were seen on the Moon by the **astronauts**, but that the story has been covered up. In the book *Elder Gods of Antiquity* (page 101), author M. Don Shorn writes, "Other anomalies include what appears to be 'domes' formed in discernible clusters, not in random patterns of natural distribution. Such domes are large, gently sloped structures, which are mostly located in or near craters. They have also been described as possible extinct volcanoes, but their true nature remains unknown."

Double Big Whack Theory

A hypothesis that a planet approximately the size of Mars collided with Earth twice in a short period of time. It was reasoned that the impacts were so intense that part of the Earth spewed out and eventually formed the Moon. The Double Big Whack Theory was eventually dismissed.

Dream

See **Charles Duke, Jr**.

Duke Jr., Charles (Charlie) Moss

1935–

Former NASA **astronaut** and Brigadier General for the US Air Force. He served as the Lunar Module pilot for ***Apollo 16*** and was the Capsule Communicator (CAPCOM) for ***Apollo 11***. He was the tenth man to walk on the Moon and at thirty-six became the youngest human to do so. While there, Duke left a photograph of his family on the lunar **surface**. It was a picture of himself, along with his wife and two sons. On the back of the picture Duke wrote: "This is the family of **astronaut** Charlie Duke from planet Earth who landed on the Moon on April 20, 1972." Duke is also known for having a pre-flight dream about his trip to the Moon that later gave him a feeling of "déjà vu. A week before *Apollo 16* headed to the Moon, Duke had a mysterious dream about meeting his twin-self during the mission. He told astronaut John Young who would be accompanying him to the lunar surface about the unusual dream. Duke relayed that he had dreamt that the two were traveling in the Lunar Rover going upwards and eventually found themselves at a ridge. In front of the ridge they saw that there were tracks.

They immediately contacted officials on Earth and obtained authorization to follow them. After obtaining the go-ahead, they travelled for a period and later located a second Lunar Rover with two individuals that were identical to themselves. The astronauts proceeded to take parts from the second Lunar Rover to later show people. In the dream, the men returned to Earth safely. Days later, after finally reaching the Moon, Duke found himself in the very same place as he was in his dream. The two men were traveling north in the Lunar Rover. Duke was steering. Their goal was the north Ray crater. As they journeyed over the lunar surface and began to traverse a hill, Duke recognized the area from his dream. There was the same grayish black, crystallized hill; the exact place that he had found himself in his dream, days earlier.

E

Earthrise Photograph

An iconic photograph taken by **William Anders** as he served aboard ***Apollo 8*** as the Lunar Module pilot. On December 24, 1968 while working under the darkness of the Moon's sky, Ander's snapped a picture of Earth that is now known as *Earthrise*. It has been hailed as "the most influential environmental photograph ever taken," by acclaimed wilderness photographer Galen Rowell. It remains the greatest picture ever taken from the **Apollo** missions.

Eclipse

See **Lunar Eclipse**.

Eden Hypothesis

The idea that the Moon was brought into Earth's orbit by advanced beings to assist mankind and create a paradise on Earth, where life would flourish and the planet would be a beautiful habitat for mankind and all creatures on it.

Edwards, Frank

1908–1967

A renowned radio broadcaster and author. Edwards enjoyed writing on subjects of mystery such as the paranormal and the supernatural and turned out books in the 1960s that are still popular today. Some of his works include *Strange World*; *Strangest of All: Authentic Stories of Fantastic Powers and Astounding Events That Science Cannot Explain;* and *The Strange World of Frank Edwards: A Psychic Investigator's Most Terrifying Encounters With the Unknown*. Edwards investigation of Soviet cosmonauts made headlines when he disclosed that the Russians had lost a minimum of five crewmen in space-related incidents. The full article appeared in *FATE* magazine in the July 1962 installment. On the subject of the Moon, Edwards wrote in, *Strange World* (page 274), "The more you study the Moon, the more you will become aware that it is an orb of mystery—a great luminous Cyclops that swings around the Earth as though it were keeping a celestial eye on human affairs." See **Cosmonauts**.

Egypt(ian)

See **Arcadians; Blair Cuspids; Farouk El Baz; Monuments; Soviet Union; Tycho Crater.**

El Baz, Farouk

1938–

Prominent Egyptian-American space scientist, geologist and selenologist. El Baz was secretary of the landing site selection committee for NASA's **Apollo program**. He worked with NASA on the **Apollo** Moon missions, lending his expertise in the areas of science and teaching the **astronauts** ways to interpret Moon observations and photography. He was later employed as a professor at Boston University, where he was the director of the Center for Remote Sensing. In a piece of trivia, to pay tribute to him, an episode of *Star Trek: The Next Generation* featured a spacecraft they named *"El-Baz."* He once stated that "not every discovery has been announced to the public," leading some to speculate that there was a great mystery on the Moon. During a period when the ***Apollo 16*** astronauts were witnessing flashes of light, El Baz commented, "There is no question about it. Not natural." El Baz also had ideas about the Moon's density. He deduced that there were caverns within the Moon that were so big that it could be considered that the Moon is hollow.

Elger, Thomas Gwyn
1836–1897

Renowned British selenographer. In 1867, Elger watched as an odd light suddenly materialized on the moon. He reported that it was there for nearly two hours before it vanished.

Endeavor, The
See ***Apollo 15***.

Endymion (Crater)

An impact **crater** where one Moon observer claims to have seen a city. Conspiracy theorists believe that there is a hologram or a type of veil or camouflage concealing the Moon in an effort to hide the truth about our **satellite**. It has been speculated that what was witnessed that day was the veil collapsing, just long enough for Moon observers to see the truth.

Enki and Ninki

Ancient Sumerian gods from the planet Nibiru. In modern times, they are recognized as **extraterrestrials** that posed as gods. **Enki** along with his sister Ninki, two of the royals, were responsible for creating humans to work as slaves to mine for gold. Enki and Ninki and their people are believed by some to have operated from the Moon.

Enoch

Biblical prophet. He was the son of Jared and the father of Methuselah. According to the Bible, Enoch lived 365 years before he was "taken" by God. To some, Enoch is viewed as an ancient astronaut. There is an account of Enoch being taken to heaven by an angel in the Pseudepigrapha. There Enoch witnessed what he referred to as angels in classes learning information and gathering knowledge about Earth. Some have speculated that this journey was really a trip to the Moon. In another account, Enoch was carried to space by two large men that had eyes like fire, who were described in the Pseudepigrapha as "the likes that had never been seen before on the Earth." These men, according to C.L. Turnage in her book *ETs are on the Moon and Mars* (page 57), gave Enoch a view of the "power of the Moon's light." When referencing this account, Turnage takes it one step further asking, "Did Enoch have the unique experience of being the first earthling to see a Nibiruin's moon base?"

Eratosthenes (Crater)

A deep impact **crater** that is named after the ancient Greek astronomer Eratosthenes of Cyrene. In his book *Our Mysterious Spaceship Moon*, author Don Wilson cites Volume 20 of the *Astronomical Register* on a peculiarity found at Eratosthenes. He states that there is "a geometric object shaped like a cross in the lunar crater Eratosthenes." Lunar researchers have observed the cross-shaped object and suggest that it is simply a trick of the eyes. Mysterious lights have been recorded in the crater. On May 4, 1936, a moon observer witnessed odd, small glowing spots on the floor of the crater. On October 4 of the same year, numerous small bright lights were seen in the same area. They were again witnessed on October 25. In 1961, on October 18, a strange starlike point of light was also observed within the crater.

Eudoxus (Crater)

A prominent impact **crater** where one of the most mysterious cases of **transient lunar phenomenon** occurred. In the late 1800s, several moon observers witnessed what looked to be a glowing cable slowly stretching across Eudoxus from west to east, until it covered the entire crater. They watched for nearly an hour and then it flickered out. It was the first and last time such a sighting was ever reported on the Moon. Additionally, on February 20, 1877, Monsieur Trouvelot, of the Meudon Observatory in France, also witnessed a mysterious light in Eudoxus. On January 29, 1882, a strange, unexplained shadow was seen in the crater. It was observed for thirty minutes. On February 27, a second shadow appeared. It was monitored for an hour and fifteen minutes. The reasons behind the mysterious light and the origin of the shadows remain a mystery.

Evans, Ronald

1933–1990

Former Apollo **astronaut** and aeronautical engineer. Evans was the Command Module pilot on ***Apollo 17***, the last **Apollo mission** to the Moon. Reportedly Evans witnessed large, unexplained flashes of light over the western part of the Moon while serving on board *Apollo 17*. In an interview in *Paranoia, The Conspiracy Reader Magazine,* NASA scientist **Farouk El Baz** commented on the situation, noting that the flashes of light were neither comets nor natural. The mysterious flashes of light were also seen by the astronauts on ***Apollo 16***. Some speculate that they were created by otherworldly beings that were in the vicinity of the Moon.

Extraterrestrial Base(s)

A center of operations originating outside of the Earth. Lunar researchers have speculated for years about the possibility of there being an **extraterrestrial** base on the Moon. Many believe that this alleged base is located on the Moon's **far side**. Researcher and author **Harold T. Wilkins** writes in his book *Flying Saucers on the Attack*, "I advance the theory that our moon has been and still probably is used as an advanced observation base, in regard to our Earth, by mysterious cosmic visitants connected with the flying saucer phenomena." There are five hypotheses as to why extraterrestrials would have a base on the Moon. 1) *Earth Observation*. The theory that extraterrestrials are monitoring the development of life on Earth and that a base on the far side could be easily hidden from humans. 2) *To Warn Earth*. In the 1950s, when the atomic bomb was first created, there was an increase in **UFO** sightings. Per some extraterrestrial **contactees**, advanced beings sought to warn the people of the Earth of the ramifications of creating such a powerful weapon. During that same period, extraterrestrials are said to have come to the Earth and visited certain governments in an effort to stop the usage of this deadly weapon. The Moon is thought to be the perfect place for these particular extraterrestrials to set up a place of observation of mankind's usage of nuclear weapons. 3) *Stopover*. Some witnesses of **UFOs** near the Moon have suggested that it may serve as a stopover point for extraterrestrials visiting Earth or even traveling elsewhere in the solar system. 4) *Mining of the Moon*. One prominent theory is that extraterrestrials are mining the Moon for minerals, which would require a base of operations. 5) *Military Intent Towards Earth*. Some claim that structures have been found on the Moon that resembles structures that would be found on a military forward base. Some claim that these buildings belong to otherworldly beings that could possibly be a threat to the Earth and her inhabitants.

Extraterrestrial(s)

A life form that originates beyond Earth and her atmosphere. There are those who suspect that extraterrestrials reside on the Moon and that they are the reason behind the various phenomena observed on the Moon over the centuries including **transient lunar phenomenon**, structures appearing and disappearing, strange noises over the **astronauts'** radios, unexplained shadows, objects moving across the Moon, **UFOs** and more. Some suggest that if there are beings on the Moon, their origins are not from Earth's stellar neighborhood. Some go as far as to suggest that the Moon is actually a foreign country, the territory of extraterrestrials and is not "Earth's Moon" as we have been taught to believe.

F

Face on the Moon

A **UFO** investigator by the name of Scott C. Warring believes that he has discovered a human face located in a crater, found on the ***Chang'e-2*** orbiter Moon map. Warring's research adds to the discussion of the possibility of **extraterrestrials** on the Moon and whether or not they are humanoid. Warring's video can be located at *UFO Sightings Daily*, *Alien Face Discovered in Chang'e 2 China Moon Map*, https://www.youtube.com/channel/UCQhbp52zX0ySztdZ5Oi1rXQ.

Faked Moon Landing

See **Moon Landing Conspiracy Theory**.

False Crater Floors

From examining some of the photographs of **craters** on the Moon's **surface**, there are lunar researchers that believe they have found false crater floors. Some maintain that if the floors are false, they may be used to hide a **base** underground or even a city beneath the lunar surface. One researcher compared the idea to a James Bond movie where a launch facility was hidden inside of a volcano with false floors. It has been speculated that there is a secret government that knows of an underground facility on the Moon. This line of thought is further fueled by the **astronauts'** mysterious comments where they spoke of people living in craters. See **Appendix 1: NASA Transcripts (*Apollo 11*).**

Far Side

The side of the Moon facing away from Earth. Only a handful of humans have seen the Moon's far side. The first photographs of the far side were on September 13, 1959 by the Soviet's *Luna 3* space probe. It is often referred to as the **dark side** of the Moon, even though it receives sunlight. Over the years there have been a number of conspiracy theories, stories and rumors about what exists on the Moon's far side. One conspiracy theory suggests that advanced beings deliberately placed the Moon in a position so that the same side always faces the Earth, so that humans cannot observe the activities on the far side. One of these activities includes **extraterrestrials** studying the Earth and her inhabitants. Another theory is that extraterrestrials are **mining** the moon. Others speculate that extraterrestrials have secret **bases** on the far side, from where they execute their reconnaissance work undetected. Some also believe that humans only seeing the far side is the extraterrestrials' way of keeping mankind in the dark about their place in the universe. One interesting theory for why the Moon only faces one position is there is a directive in place from advanced beings that prevents extraterrestrials from revealing themselves to the people of Earth until mankind is ready.

In 1959 the **Soviet Union** sent the space probe *Luna 3* to the Moon. *Luna 3* photographed the far side and for the first time, humans could see the Moon's backside. It was found to look very different from the nearside (the side facing Earth). From the photographs taken by *Luna 3*, we learned that the far side is extremely rugged, contains a large number of impact craters and relatively few maria. Don Wilson, author of *Our Mysterious Spaceship Moon*, writes, "About a third of the Moon facing the Earth is made up of these maria while there are few on the 'dark side,' and no-one can explain why the two sides are so different." Additionally, there are claims of ancient structures and **ruins** having been found on the far side. According to some conspiracy theorists, information on these remnants was withheld from the general public. In addition, there are some **contactees** that claim that the far side of the moon is inhabited.

Fastwalkers

Unidentified flying objects that advance from deep space, enter the vicinity of Earth and move about in ways unidentifiable with regular, recognizable aircraft. The word "fastwalker" was reportedly created by the North American Air Defense Command (NORAD). The strange objects, often associated with ufology, have also been seen traveling rapidly across the Moon and have been caught on video. Purportedly, NORAD watches these objects from an underground base located in Cheyenne Mountain in El Paso County, Colorado.

Female Extraterrestrial

See **Mona Lisa**.

Fireflies

A name given to the small, lighted **objects** that surrounded astronaut **John Glenn** while on the *Mercury-Atlas 6* space mission. During the period that Glenn was orbiting the Earth, he had an encounter with thousands of lights that surrounded the craft. He described them as "little stars" that came by in "showers." They were luminescent, small (less than a sixteenth of an inch,) approximately seven to eight feet apart and were traveling at a slow rate of around five miles an hour. Amazingly, the same type of phenomena was also described by **contactee** George **Adamski** in the early 1950s. In his book *Flying Saucers Have Landed*, Adamski writes of going on board an **extraterrestrial** mothership that was as large as a traveling city. At 50,000 miles from Earth, Adamski looked out into space through one of the windows. He described space as being totally dark. Yet in the midst of the blackness, he saw what looked to be billions upon billions of small lights moving about. He observed that they were in a variety of colors and flickering and moving about in many directions. Adamski likened the phenomena to watching a cosmic show of fireworks.

First Contact

A popular term used in science fiction that has crossed over into the mainstream referencing a time when mankind will first meet **extraterrestrial**s. There are those who believe that first contact was made during ***Apollo 11***, the **Apollo program's** first mission to land men on the Moon. Some refer to a story of **Neil Armstrong** and **Buzz Aldrin** allegedly seeing extraterrestrial **spaceships** sitting on the rim of a crater; they were described as being large and menacing. The extraterrestrials, as the story goes, warned the **astronauts** off the Moon. There is also a tale that has circulated, where one of the astronauts (from *Apollo 11*) purportedly saw an **extraterrestrial.** Some believe that the astronauts were forbidden to speak of their experiences with extraterrestrials, which would have essentially been first contact. See **Warned off the Moon**.

First Moon Photograph

The first photograph of the Moon was taken on March 23, 1840 by American photographer John William Draper (1811–1882).

First Person on the Moon

See **First Step**.

First Step

The first step by man on the Moon's **surface** was made by astronaut **Neil Armstrong** of the ***Apollo 11*** mission in 1969. It has been widely reported that Armstrong stepped out of the Lunar Module and onto the Moon's surface, but it was actually a large jump onto the surface. This was because the Lunar Module's shock absorbers had not compressed during the landing.

Fission Theory

Originally proposed by George Darwin (son of Charles Darwin), the Fission Theory is the hypothesis that in the beginning, a soft, spinning molten Earth revolved so fast that it became distorted and in the process a portion tore off and was flung into space, eventually forming the Moon.

Flag Conspiracy Theory

On the first mission to the Moon (***Apollo 11***) one of the assignments of the astronauts, was to place a flag on the Moon's **surface**. On July 20, 1969, **Neil Armstrong** and **Buzz Aldrin** carried out that task. The following Moon landings followed suit. The first flag to be placed on the Moon was blown away when the **astronauts** blasted off in the Lunar Model to rejoin the Command Module. In a video of the astronauts of ***Apollo 15*** placing an American flag on the Moon, the flag can be seen waving. The flag waving on the Moon has added to the conspiracy theory that the Americans never went to the Moon.

Flammarion, Nicolas Camille
1842–1925

A noted astronomer from France. He was the author of several popular science fiction novels and owner of the magazine *L'Astronomie*. Flammarion had a personal observatory in the township of Juvisy-sur-Orge, France. Flammarion postulated that almost all of the planets within our solar system were populated. He viewed the Moon as an inhabited world complete with an **atmosphere**, water, vegetation and a population. It was Flammarion that came up with the term **Selenites**, used to identify Moon inhabitants. Studying Flammarion's teachings, researchers have questioned whether he observed something different from what we see today—something mysterious that may have since disappeared.

Flashes of Light

A type of **transient lunar phenomena** (TLP). Flashes of light have been seen on the Moon by astronomers for hundreds of years. NASA's **Apollo** astronauts witnessed them at close range, but could not locate the source. Astronaut **Harrison Schmitt** reported seeing a flash of light on the Moon's **surface**, stating, "It was a bright little flash near the crater at the north edge of **Grimaldi**." Astronaut **Ronald Evans**, who was the Command Module pilot for ***Apollo 17***, witnessed a flash of light close to the rim of Mare Orientale, exclaiming, "Hey! You know, you will never believe it. I'm right over the edge of Orientale. I just looked down and saw a light flash myself." During the ***Apollo 16*** mission, astronaut **Kenneth Mattingly** stated: "Another strange sight over here. It looks...a flashing light." **Rudolph Lippert** of the British Astronomical Association, while observing the Moon, saw a yellowish orange light that flashed. Flashes of light were also seen in **Mare Crisium**. No one has an explanation for these odd lights. Some wonder if these flashes of light are deliberately shown to attract the people of Earth's attention. See **Appendix 2: Anomalous Lights; Blue Lights; William Herschel; Luminosities; Orbs.**

Flood
See **Great Flood**.

Footprint(s)

Neil Armstrong is famous for being the

first man to step foot on the Moon. He made the historic step on July 20, 1969. He stepped onto the Moon's **surface** with his left foot. That footprint left on the moon measures 13 by 6 inches, which were the measurements of his boot. It is estimated that the footprint from the **first step** and the others from additional **astronauts** that walked on the Moon, will last on the Moon's **surface** for approximately 10 million years. According to one mysterious and fantastical tale**,** ***Apollo 11*** astronauts **Neil Armstrong** and **Buzz Aldrin** saw strange footprints when walking on the Moon. Aldrin is even said to have taken pictures of the mysterious prints. Eventually, according to one story, not only did they see footprints, but they saw the beings that had made them.

Foreign Country

A term used to imply that the Moon is a nation of its own, where there are extraterrestrials living that have no connection to the Earth. The belief that the Moon is inhabited and belongs to someone else dates back centuries. There were many philosophers in the past who believed that there was life on the Moon. Among these were **Anaxagoras of Clazomenae, Aristotle**, **Lucian of Samosata**, **Plutarch**, Pythagoras and Xenophanes. The noted natural philosopher and author John Wilkins in the mid 1600s expressed his ideas of there being life on the Moon in his work *Discovery of a New World in the Moon,* stating, "It is probable there may be inhabitants in this other World, but of what kind they are is uncertain." In more recent times, people also espouse the idea of the Moon being populated by extraterrestrial beings. Some believe that these lunar people are the descendants of advanced beings that originally brought the Moon into our solar system, purposely placing it in Earth's orbit. In an article written by author Steve Omar titled "Moon and Mars: The Moon is a Foreign Nation," Omar ponders, "Could it be that the Moon is a foreign country and someone else's property and the Moon's government does not want us coming up and invading their territory with our nuclear weapons, pollution, unwelcome military facilities, diseases, litter, mining exploitation and historicly proven record of foreign imperialism?"

Forward Base
See **John Brandenburg**.

Four Blood Moons
See **Tetrad**.

Four Moons
See **Hans Hoerbiger**.

France
See **Appendix 2: Anomalous Lights (1920 and 1930); Robert Charroux; Domes; Camille Flammarion; Eudoxus Crater; Pierre-Simon Laplace; Thierry Speth**.

Full Moon

The period when the Moon appears completely round and luminescent. Mysterious events often occur during a full moon. There is a perception among some doctors and also nurses, that more people visit medical facilities for gastrointestinal bleeding, or hemorrhaging, during a full moon. Additionally, some hospitals and blood banks have noticed a higher demand for blood during this time. A study was run in one facility for the purpose of determining whether the number of people being seen for this problem was higher on full moon days. Although the conclusion from the study was that the full moon hypothesis is true, the majority of people remain skeptical of this idea. The human body is made up of 60 percent water. Some believe that since the Moon has an influence on the oceans' tides, that it also has an effect on the human body and the brain. A full moon is thought to affect emotions, causing heightened feelings, passions and sometimes negative thoughts. It is believed by some to be when a person is most likely to carry out a crazy and extreme deed. Accoringly, there is a theory that there is a link between the full moon and criminal activity. Authorities in various areas have concluded that when there is a full moon, there is more violence and crime. While this view was accepted in years past, scientists today often deny it to be true. See **Human Body.**

Fusion Reactor
See **Aristarchus Crater; Copernicus Crater**.

G

Gamma

The third letter of the Greek alphabet. The letter gamma is associated with a Moon mystery. Near the Moon's **Littrow crater**, there are symbols that appear to be the letter gamma. In his book *Our Mysterious Spaceship Moon*, author Don Wilson cites the *Astronomical Register,* Volume 20, where it reports, "on the floor the crater Littrow are seven spots in the shape of the Greek capital Gamma." There is a belief that Earth and the Moon were connected during ancient times. It has been speculated that the symbol Gamma was possibly put there by an ancient people that once existed on the Moon. Don Wilson concludes by suggesting that they are from **extraterrestrials** that once used the Moon as a **base** of operations.

Gagarin, Yuri Alekseyevich
1934-1968

A former Soviet pilot and cosmonaut. He has the distinction of being the first man in space. In the Soviet's desire to reach the Moon, they began sending humans on missions to outer space. Gagarin was the first to accomplish that goal. He orbited the Earth on April 12, 1961. and returned to Earth successfully and "relatively unharmed."

Gassendi (Crater)

A large impact crater. Mysterious lights and activity have been witnessed and also monitored in the crater. In his book *Our Mysterious Spaceship Moon*, author Don Wilson cites the *Astronomical Register,* Volume 20, where it reports, "In the lunar **crater** Gassendi are angular lines." The lines are believed to be unnatural and are thought by moon researchers to have been left there by otherworldly beings that were working on the Moon, sometime in the ancient past. Famed astronmer **Patrick Moore** witnessed a peculiar reddish glowing light on the Gassendi crater, describing it as being, "a color phenomena." This strange occurrence was said to have been witnessed by two other astronomers, including P. Satory and T. Moseley. Other strange and mysterious activity has been reported in and around the crater. A compilation follows.

1889, May 11, a strange black spot was seen in the area of the rim.

1951, May 17, a bright spot was seen and monitored in the crater for a short period of time.

1966, April 12, a sudden flash of red light appeared, settling into a ruddy haze near the northwest wall. It was observed for eighteen minutes.

1966, April 30 to May 2, red glowing lights were observed.

1966, September 2, ruddy-colored patches were witnessed. They were observed for three hours.

1966, September 25, reddish colored areas were seen. They were monitored for 30 minutes.

1966, October 25, red blinking lights were noticed on the crater's north wall.

1966, December 27, dim blinking lights were seen on the southwest and northwest areas of the surface for thirty-five minutes.

1967, March 22 and 23, red blinking lights

were witnessed.

1967, May 20, an elongated set of flashing lights were seen in the southwest part of the floor.

1967, June 18, dim ruddy-colored areas were noticed near the northwest and Southwest walls.

Gateway of the Sun

See **Sun Gate at Tiahuanaco**.

Gauss, Johann Carl Friedrich

1777–1855

German mathematician, physicist and director of the Gottingen Observatory. Gauss once proposed a method of signaling to possible residents of the Moon.

Gemini 1

*Gemini o*ne of NASA's **Project Gemini** spaceflight program was launched on April 8, 1964. Project Gemini was a precursor to the future Apollo missions to the Moon and a bridge between the Mercury and **Apollo programs**. *Gemini o*ne was the first test flight. It was unmanned. Its function was to test equipment and mission procedures to train **astronauts** for future **Apollo missions** to the Moon. According to **Donald Keyhoe**, who was once the director of the *National Investigation Committee on Aerial Phenomena (NICAP),* during the flight, *Gemini o*ne detected four **UFO**s that followed the craft into orbit. Scientists were following the Gemini flight on radar and they too picked up the mysterious **objects.** The UFOs followed *Gemini o*ne for one complete orbit and then quickly flew away. No explanation was ever given as to what the mysterious objects were or where they came from. UFO researchers speculate that mankind is being monitored by otherworldly beings, especially when it comes to defense programs and space missions. It is believed that these beings were well aware that the Gemini program was running tests and developing procedures for future Moon expeditions and were there to watch.

Gemini 4

The second spaceflight of **Project Gemini** and a continuation in the space agenda to land men on the Moon. It was the program's first crewed mission. It launched on June 3, 1965. The crew consisted of Command Pilot **James McDivitt** and Pilot **Edward White**. *Gemini 4* was tasked with operating an extravehicular activity (EVA) and practicing orbital maneuvers. On the *Gemini 4* mission, McDivitt and White observed a luminous, **cylindrical-shaped** object. The object was in close proximity to their position. It was so close that the **astronauts** worried about a possible collision. They reportedly went so far as to plan to take evasive actions if it came any closer to their spacecraft. This sighting occurred as they travelled over Hawaii. The object was described as a glimmering, silver-colored, cylindrical-shaped **UFO** with elongated appendages stretching out from it, heading towards the craft. As McDivitt watched the object approach, he took a picture. Although the film was examined by NASA, the object was never identified and remains a mystery.

Gemini 7

Gemini 7 was the fourth mission under the **Project Gemini** spaceflight program to hold a crew. It was a continuation of preparation to future trips to the Moon. *Gemini 7* launched on December 4, 1965. The objective was to prove that human beings could survive in a weightless environment for 14 days. The crew consisted of **Frank Borman** and **James Lovell**. *Gemini 7* was in space for 14 days and made 206 orbits. It was during the second orbit that Borman and Lovell encountered the unexpected. Lovell contacted Mission Control in Cape Kennedy to inform them that they were seeing mysterious **objects** outside of their capsule. When asked if they were seeing the booster rocket, Lovell responded that they saw not only the booster rocket, but several other objects as well. See **Appendix 1: NASA Transcripts (*Gemini 7*).**

Gemini 9A

The seventh of the Gemini missions. It launched on June 3, 1966. Astronauts **Tom Stafford** and **Eugene Cernan** served aboard. It was tasked with running necessary tests for future missions to the Moon. According to sources, the **astronauts** of *Gemini 9A* saw that they were being followed by several **UFOs**. Ground control also spotted the objects and began to monitor them. What it was that the astronauts saw following them remains a mystery.

Gemini 10

The eighth of the **Project Gemini** space operations. It launched on July 18, 1966. Members of the crew included Command Pilot John Young and Pilot Michael Collins. During their mission, the **astronauts** reported that two "bright objects" were following their spacecraft. Eventually, the objects sped out of sight. Later, the astronauts reported strange objects following them a second time. The astronaut's best guess as to what the objects could be was some type of satellites. See **Appendix 1: NASA Transcripts (*Gemini 10*).**

Gemini Program

See **Project Gemini**.

Genesis Bean

A small fragment of green glass shaped like a lima bean that was found in the lunar soil **samples**. It was dubbed, "the Genesis bean." The lunar samples had been collected by the ***Apollo 14*** astronauts during their journey to the Moon.

Genesis Rock

A moon rock containing traces of water that was found and brought back by ***Apollo 15*** astronauts for study. **Astronauts** James Irwin and **David Scott** came across the now-famous rock sample on August 1, 1971 while at the Spur **crater**. It is estimated to be nearly 4.5 billion years old, among the olest moon rocks sampled. The **Genesis Rock** is said to be the most important find of the *Apollo 15* mission. It is housed in the Lunar Sample Laboratory Facility in Houston, Texas.

Geo-Dome, The

An anomalous construction located on the rim of the **Tycho crater**. It is featured in author and researcher **Mike Bara's** book *Ancient Aliens on the Moon* (page 187). Bara describes it as "roughly hexagonal in shape," and says it "seems to jut from the hillside as if it were the tip of a much larger sub surface object." The geo-dome is also described as having a roof, a wall and windows.

The hexagonal geo-dome on the Moon.

Germans

A mysterious tale states that allegedly the Germans went to the Moon in the 1940s, years before the Soviet Union and the United States. They purportedly had advanced technology to help obtain this goal. Additionally, they are said to have placed a **base** there. It is said that they obtained knowledge from recovered **UFOs** that had crashed. Others contend that the Germans had contact with **extraterrestrials** and information was obtained directly from them.

Giant Impact Theory

The Giant Impact Theory (also known as the Big Whack Theory, the Theia Impact and the Big Splash Theory) is the most popular and generally recognized hypothesis as to how the Moon was created. It was proposed in 1975 by two PSI senior scientists, Dr. William K. Hartmann and Dr. Donald R. Davis. The two wrote a paper on the subject that appeared in the scientific journal, *Icarus*. Their theory suggests that the Moon was formed when a large planetary object thought to be Theia (a hypothesized ancient planetary body) smashed into Earth. Theia was around the same size as the planet Mars. This giant impact is believed to have caused huge fragments of the Earth to break off, sending them flying into space and eventually creating the Moon. The larger pieces are believed to have possibly broken off from the Earth's mantle. The time frame of this collision has been placed at nearly 4.51 billion years ago, during the Earth's Hadean period.

Glass Domes

See **Domes**.

Glass-Crystal Structure

In an ***Apollo 16*** transcript, astronaut **Charles Duke** talks about what sounds like a mysteriously beautiful, part glass, part crystal-like structure, located at the Descartes site. He states, "Right out there... the blue one I described from the Lunar Module window is colored, because it is glass coated. But, underneath the glass it is crystalline... the same texture as the Genesis rock... dead on my mark." Some believe that the **astronauts** at that moment were looking at an artificial construction.

Glenn Jr., John Herschel

1921–2016

Former NASA astronaut, Marine Corps pilot and a Senator from Ohio. Glenn served aboard the *Mercury-Atlas 6*, a part of the Project Mercury program. Project Mercury was a prelude to the future **Project Gemini** and ultimately the **Apollo Program** where there would be landing operations on the lunar **surface**. He also served aboard the STS-95 Space Shuttle mission. He was a recipient of the *NASA Distinguished Service Medal* in 1962 and the *Congressional Space Medal of Honor* in 1978. Glenn was the first United States **astronaut** to travel around the Earth, orbiting it three times (in low orbit). He also had the distinction of being one of the Mercury Seven, the original test pilots chosen in 1959 by NASA to become the nation's first astronauts. During the Mercury mission, while he was orbiting the Earth, Glenn had an encounter with thousands of odd and perplexing lights that surrounded his spacecraft. He described them as "particles" and "little stars" that came by in "showers." Glenn said they were luminescent, small (less than a sixteenth of an inch), approximately seven to eight feet apart and were traveling at a slow rate of three to four miles per hour, slower even than the capsule he was traveling in. There have been theories brought forth about what it was that Glenn saw including ice particles and radiation from the sun. The small lights were later dubbed, "fireflies." In another account, Glenn had a **UFO** experience, again while orbiting the Earth. Glenn witnessed three UFOs quickly coming towards his spacecraft. The UFOs caught up to his craft, then passed it and eventually disappeared. See **Appendix 1: NASA Transcripts (*Mercury-Atlas 6*).**

Gleti

A moon goddess from the beliefs of the people of Benin, West Africa. Gleti presides over the Moon and is mother to the gletivi, a group of cosmological gods and goddesses that later became stars in the heavens.

Gondwana (Gondwanaland)

Ancient **Tibetan** writings point to the lost continent of Gondwana as having existed during the mysterious pre-lunar period.

Goodacre, Walter

1856–1938

Businessman and amateur British astronomer. Goodacre was director of the Lunar Section of the British Astronomical Association from 1897-1937. He was president from 1922-1924. In 1910, he published a 77-inch diameter hand drawn map of the Moon. In 1931, he published a larger book of maps of the Moon's **surface** with descriptions of its features. In 1931 Goodacre observed what appeared to be an unnatural shimmering **blue light** on a wall of the **Aristarchus crater** in the early morning hours. He reported his unusual sighting to the British Astronomical Association.

Goodwill Messages

See ***Apollo 11*** **Goodwill Messages**.

Gordon Jr., Richard Frances

1929–2017

Former NASA **astronaut** and American naval officer. He was the pilot for the *Gemini 11* mission and the Command Module pilot for ***Apollo 12***. In September 1966, while aboard *Gemini 11,* Gordon and fellow crewman **Charles Pete Conrad** spotted a **UFO,** whose color was described as yellow-orange, located approximately six miles away from their position. The **astronauts** attempted to capture the strange object on film, but were unsuccessful in acquiring a clear image.

Grave

See **Eugene Shoemaker**.

Gravity

The Moon's gravity is one-sixth (1/6) that of the Earth's.

Grays, The

See **Zetas**.

Great Flood

A story of a great worldwide flood can be found in tales from various cultures around the world. The most well-known version in the West is the biblical story of **Noah's ark**, in which God saved a limited number of humans and animals on a large, specialized boat. God had given directions to a man named Noah on how to build it. The account states at Genesis 7:17-20, "The flood continued upon the earth for forty days. As the waters increased, they lifted the ark, so that it rose above the earth. The waters swelled and increased greatly on the earth, but the ark floated on the surface of the waters. Higher and higher on the earth the waters swelled, until all the highest mountains under the heavens were submerged. The waters swelled fifteen cubits higher than the submerged mountains." (*New American Bible*) Another story is the Sumerian flood myth; written on a tablet, it tells of a large boat that was caught up in a horrendous storm that lasted seven days and seven nights. The great flood is believed by some lunar researchers to have been caused by the Moon entering Earth's orbit around twelve thousand years ago. The Moon's entrance is believed to have caused great changes upon what would have been a tumultuous planet, existing without a **satellite** to stabilize it.

Great Moon Hoax

A series of six sensational articles that were published in New York's *The Sun* newspaper in 1835 that had people convinced that an extraordinary civilization had been discovered on the Moon. The supposed purpose of the story was to announce the finding of life on the Moon by famed astronomer **John Herschel**. The story was allegedly written by Richard Adams Lock (a reporter and associate of Herschel) under the pseudonym of Dr. Andrew Grant. The headline of the story was Great Astronomical Discoveries Lately Made By Sir John Herschel, L.L.D.F.R.S. &c., From Supplement to the Edinburgh Journal of Science. These newly found discoveries were said to have been made by Herschel, as he worked at the Cape of Good Hope, South Africa. Herschel had traveled and set up a base there for his studies. The article spoke of Herschel finding life on the Moon, spotting such amazing creatures as bat-like humans, various animals (including bison, goats, unicorns and bi-pedal beavers). He also noted the geography which included large **craters**, trees, oceans, beaches, rivers, rich **vegetation** and large amethyst crystals. The story came to an end when it was announced that the telescope and the observatory had been destroyed, due to a fire caused by a burning lens set aflame by the Sun's heat.

Greaves (Crater)

An undersized impact crater situated near the southwestern rim of **Mare Crisium**. Greaves has a spherical formation that has a slight inner floor in the middle of slanting internal walls. The **crater** has become a source of close observation in recent years, due to a puzzling object located near it. To some lunar researchers, the mysterious object looks to be a "satellite dish."

Green Moon Hoax

A hoax perpetrated in the mid-2010s stating that the moon would turn green and does so every 420 years. The rampant use of social media assisted in the spread of this hoax. People from around the world expected to see a green moon. The green moon hoax first appeared in 2016, when a member of the social networking website Facebook posted a picture of a green moon, which they claimed could be seen in the following spring. Many people that believed there would be a green moon shared the information with friends and waited for the spectacular sight that would only occur once in their lifetime. It is believed that the green moon hoax was inspired by real moon phenomena such as the black moon, blood moon, blue moon, orange moon and red moon.

Greenacre, James Clark

1914–1994

Astronomer and a United States Air Force Aeronautical Chart and Information Center (ACIC) cartographer. He later became a United States Air Force Lunar Cartographer-Observer for NASA's **Apollo Program** at Lowell Observatory in Flagstaff, Arizona. On the evening of October 29, 1963, Greenacre and his associate **Edward Barr** were working at Lowell Observatory. Their job was to map the **Aristarchus crater** which is twenty-seven miles in diameter. Both Greenacre and Barr knew the crater and the area around it extremely well. Greenacre looked through the observatory's twenty-four inch telescope and was astonished to see an intensely glowing, reddish-orange and pink light on the Moon's **surface**, located near Cobra's Head and the southwest interior rim of the **Aristarchus crater**. Greenacre was said to be shocked to see the strange phenomena. He described the scenario as to be if "looking into a large gem ruby." This strange **transient lunar phenomenon** (TLP) sighting is said to have sparked more interest in observing TLPs on the Moon. In another account, Greenacre saw a second mysterious TLP on November 28, 1963, as he worked alongside Barr and ACIC scientific illustrator Fred Dungan and Lowell Observatory Director John S. Hall. Greenacre's sightings were reported in the December 1963 issue of *Sky and Telescope Magazine* and in the 1964 USAF report, ***Lunar Color Phenomena, ACIC Technical Paper No. 12,*** as well as at various lunar symposiums. No conclusion was ever made as to the cause of this spectacular light phenomena.

Greer, Steven
See **Disclosure Project**.

Greys, The
See **Zetas**.

Grimaldi (Crater)

An area where mysterious light activity has been seen. During the ***Apollo 17*** mission, astronaut **Harrison Schmitt** witnessed a **transient lunar phenomenon** while studying the lunar **surface**. Schmitt was heard to exclaim, "I just saw a flash on the lunar surface!! Location: north of Grimaldi. It was a bright little flash near the **crater** at the north edge of Grimaldi." Grimaldi has been the topic of other mysterious activity as well. From March 29 to 30, 1789, two glimmering spots of light were seen on the eastern edge the crater. On June 24, 1839, what resembled a grayish-colored mist was observed by **Franz Gruithuisen**. During June 25, 1964, a bright flash was seen towards the lip the crater during a **lunar eclipse**. On the day of July 7, 1965, sa econd streak of light was witnessed in the crater.

Gruithuisen, Franz von Paula
1774–1852

A professor of astronomy and a physician. Gruithuisen is most noted for his studies and discoveries involving the Moon. He was the first to propose that the Moon's **craters** were the result of meteorites hitting the lunar floor (known as the Meteoric Impact Theory of Lunar Cratering). In July 1821, he saw brilliant, small flashing lights on the **surface** of the moon. Gruithuisen believed that there was a civilization on the Moon. His studies of the Moon's surface convinced him that he had discovered a city there. He called the city Wallwerk. The city was in a rugged area of the Moon, located north of the Schroeter crater. The perception of a city by Gruithuisen is thought to have come from the area's linear ridges that have been described as having a fishbone-like design and geometrical shapes. Today, it is believed that he mistook these lines for roads, walls and waterways. Gruithuisen also maintained that he had seen assorted hues of color on the Moon. He associated these colors with an **atmosphere** and **vegetation**, which he believed existed on the Moon. It has been stated that due to the limited telescope capabilities of his time, Gruithuisen could have easily believed that, when looking at the various shapes and images, he was witnessing a civilization on the Moon. In 1824, Gruithuisen went public with his findings publishing a paper titled "Discovery of Many Distinct Traces of Lunar Inhabitants, Especially of One of Their Colossal Buildings." How Gruithuisen was making his conclusions about a city on the Moon was a mystery to his peers. In the end, they were not convinced that his ideas were correct. His discoveries were later refuted as more powerful telescopes became available.

Gruithuisen's city on the Moon.

H

Hadley Rille

A large valley that spans 70 miles in length and has been dubbed by researchers the "Grand Canyon of the Moon." When the ***Apollo 15*** astronauts visited Hadley Rille in 1971, they found what is now referred to as the Hadley Rille meteorite. It was the second meteorite to be discovered on a solar system body other than the Earth.

Halley, Edmond
1656–1724

English astronomer most noted for discovering Halley's Comet. Halley was also one of the original people to apply mathematics in explaining the Moon's orbital movements. In March 1676, while studying the Moon, Halley spotted a large unidentified object near the orb. Halley reported that he had seen a "Vast body apparently bigger than the Moon." In addition, he added that there were sounds coming from the **UFO**. He described the noise stating that it was, "Like the rattling of a great cart over stones." After monitoring the distance of the object, he determined that it was traveling at more than 9,600 miles per hour.

Halo

A round disk that appears to encircle the Moon (or other lighted body) that is caused by the refraction of light through ice crystals in the atmosphere. Moon halos can take on a variety of forms. They can be perfect spheres or elliptical and can be single rings or multiple rings. On occasion there is a spectacular, colorful moonbow (lunar rainbow). In *Reader's Digest's Mysterious of the Unexplained* (page 250), there is an account of an unusual halo around the Moon. According to Lewis Evans, on January 21, 1913, while traveling aboard a ship off the coast of equatorial Africa, he observed a square-shaped halo around the Moon. It was described as "three moon diameters on a side with one corner down toward the horizon." States Lewis of that evening: "I pointed out the halo to some of the ship's officers, none of whom seems to have seen one like it before." *Mysterious of the Unexplained* states, "No existing optical theories can account for the square halo." The account was originally, found in the *Quarterly Journal of the Royal Meteorological Society*, 39:154, April 1913.

Halstead, Frank
1883–1967

Astronomer and former curator at the Darling Observatory in Duluth, Minnesota. On July 6, 1954, Halstead along with several others witnessed a straight black line in Piccolomini (a prominent impact crater found in the southeastern region of the Moon). This weird anomaly was not something that had been seen in the vicinity before. The phenomena was also seen and confirmed by other astronomers.

Harris, F. B.
See **Objects Crossing the Moon**.

Harrison, Henry
See **Appendix 2: Anomalous Lights (1877).**

Hecate
(Also Hekate)

A goddess of the Moon in Greek mythology. She is closely related to the waning moon, mysticism and mysteries. Her parents were the Titans Perses and Asteria. It was from them that she received her authority over the dark moon. She wears a **crescent moon** on her head.

Helicon (Crater)

An impact crater situated close to the northern area of Mare Imbrium. On March

13, 1788, French astronomer Nicolas-Antoine Nouet (1740-1811) gave a mysterious account of what he had seen at the Helicon crater. He reported that he had observed a light shining as brilliantly as a sixth magnitude star. He described it as being similar to a lunar volcano.

Helium-3

Helium-3 (He-3) is a stable (non-radioactive) isotope of Helium (He). While Helium has two protons and two neutrons, He-3 has two protons and one neutron. Although it is theoretical at this point, He-3 has the potential to become a source of fusion energy in the future since it does not create radioactive byproducts. He-3 can be found on Earth in small quantities. The most abundant sources of He-3 are the gas giants in our solar system, followed by the Moon (in its regolith). Several space agencies around the world (US, Russia, China) have been contemplating mining the moon for its He-3. However, to do this, great amounts of the regolith have to be processed to produce a small amount of He-3 (roughly 150 million to one ratio). There are mysterious tales of **astronauts** seeing mining equipment on the Moon. According to sources, a picture of the **Humboldt crater** allegedly shows an excavating operation. There is a theory that otherworldly beings may be mining on the Moon for Helium-3.

Heng-O

A moon goddess in ancient Chinese beliefs. She is the progenitor of the twelve moons and ten suns.

Herschel, Frederick William

1738–1822

Prominent German-British astronomer and composer. Herschel has been hailed as the "greatest Moon observer in the history of the science." His sister was another well-known astronomer named Caroline Herschel. His son was the famed astronomer **John Herschel**. Herschel was the creator of sidereal astronomy for the systematic observation of the heavens. He is also known for discovering Uranus on March 13, 1781. He also postulated that nebulae are made up of stars. In addition, Herschel created a theory of stellar evolution. He became the first president of the Royal Astronomical Society in 1820. There is an impact **crater** named after Herschel on the Moon (along with two others: C. Herschel after his sister, Caroline; and J. Herschel after his son, John). There are also craters named after him on Mars and on Mimas (a Moon of Saturn).

During his many years of studying the Moon, Herschel reported seeing numerous mysterious lights on and around the Moon. In one report, Herschel witnessed as many as one hundred and fifty lights in various areas of the Moon, during a **lunar eclipse**. In his book *Flying Saucers on the Attack* (page 217), John Wilkins writes about the observation by Herschel stating, "The famous astronomer Frederick William Herschel, was looking through a 20-foot reflector telescope, on October 2, 1790, when he saw, in time of total eclipse of the moon, many bright and luminous points, small and round." Herschel also reported seeing numerous lights mysteriously traveling above the Moon throughout the years. On August 18, 1787, he saw small, gleaming round patches on the Moon that resembled, in his words, "slowly burning charcoal thinly covered with ashes." In November of 1821, he reported seeing peculiar lights three times, one after the other. See **Appendix 2: Anomalous Lights (1783, 1787, 1821).**

Herschel, John

1792-1871

One of England's most distinguished astronomers. He was the son of noted astronomer **William Herschel.** It was Herschel that devised the Julian day system used mainly by astronomers. Hershel discovered the seven moons of Saturn and the four moons of Uranus, naming them in the process. During his long career, he was the recipient of several awards for his work in reevaluating the double stars and nebulae as catalogued by his father. He received the prestigious Gold Medal of the Royal Astronomical Society. He also won the Lalande Medal of the French Academy of Sciences (for scientific advances in astronomy) and the Copley Medal for his mathematical contributions to their transactions. The J. Hershel crater on the Moon honors his accomplishments in the field of astronomy. In 1835, a series of six articles were published in New York's *The Sun* newspaper that had people convinced that a civilization had been discovered on the Moon by Herschel. It was allegedly written by Richard Adams Lock (a reporter and associate of Herschel) under the pseudonym of Dr. Andrew Grant. The headline of the story was Great Astronomical Discoveries Lately Made By Sir John Herschel, L.L.D.F.R.S. &c., From Supplement to the Edinburgh Journal of Science. This newly found civilization was said to have been observed by Herschel as he worked at the Cape of Good Hope, South Africa. Herschel had set up a base there for his cosmological work. The article spoke of Herschel finding life on the Moon and seeing unusual creatures and various animals. He also noted the geography which ranged from craters to bodies of water and vegetation. The story came to an end when it was made known that both the telescope and the observatory in which Herschel had been working were destroyed in a fire. The series of articles from *The Sun* are now known as the **Great Moon Hoax**.

Hevelius, Johannes
1611–1687

A famed Polish intellectual and astronomer. His observatory was known as Stellaeburgum Starenburg, which means "the Star Castle," built in 1641. His influence can be found in modern day astronomy. He was the founder of lunar topography. In life and work Hevelius turned his attention to his love of the cosmos. He was an expert in creating and assembling telescopes and observing and recording his findings of the universe. He used these skills to create a compendium of 1,564 stars, the most detailed list of stars of the period. Additionally, he created an extensive celestial atlas. There he introduced many of the constellations we know today. Due to the limitations of the telescope during that period, Hevelius engaged in a long, slow four-year study of the Moon, as he was unable to view the entire **surface** at once. In 1647 he published an atlas of the Moon titled *Selenographia*. It included a comprehensive map of the lunar surface and featured the Moon's landscape and characteristics. Additionally, Hevelius argued that the Moon had bodies of water and that it was inhabited by **Selenites**, as he referred to them, who lived on the land. His work was honored by having a Moon **crater** named after him. He also studed some of the lunar mountains, of which he was able to measure the height by examining the shadows shown on the lunar floor. Hevelius put names to these mountains, peaks and valleys which are still used today. A hundred years later, astronomers were still using his work in their studies to learn even more about our mysterious Moon. ***The Chronological Catalog of Reported Lunar Events*** lists Hevelius as witnessing something strange on the Moon in 1650. He described it as a "red hill" near the **Aristarchus crater**.

Highlands

The lighter areas of the Moon's **surface** (the darker areas are referred to as **maria**).

Hill, Howard
(Unknown birth and death dates)

British Astronomer. In his Moon observations from the years 1950 to 1987, Hill located incredible lunar formations. These included enormous walls near the Hypatia Rille and constructions resembling pyramids in the Moretus **crater**. In his book *Extraterrestrial*

Archaeology author David Hatcher Childress writes, "At various times Hill believed he saw a central staircase and possible sand-dune movement around the pyramid."

Hina

In Polynesian mythology, Hina is the goddess of the Moon. She created the Moon which she governs, guards and resides upon. According to legend, Hina began her life on the Earth. She is said to have grown weary of toiling for her brother and traveled to the Moon to exist in tranquility.

Hippolytus of Rome

160 AD–236 AD

Noted theologian and ecclesiastical writer of the third century. In his work *Refutatio Omnium Haeresium V. ii.*, Hippolytus wrote about a time before the Earth had a moon: "Arcadia brought forth Pelasgus (the first man), he was older than the Moon." His words have been used by some as proof that Earth did not always have a Moon, adding to the mystery of the **origin** of the Moon.

Hoagland, Richard

1945–

American author, researcher, lecturer and former science consultant to NASA. Hoagland also worked as a science consultant for broadcast journalist **Walter Cronkite** and *CBS News* during the **Apollo** missions to the Moon. Hoagland alleges that NASA is retaining information about an ancient civilization that once existed on the Moon. His books include: *Dark Mission: The Secret History of NASA* (co-authored with Mike Bara) and *The Monuments of Mars: A City on the Edge of Forever.*

Hoax(es)

See ***Apollo 20*****; Great Moon Hoax; Green Moon Hoax; Moon Landing Hoax**.

Hoerbiger, Hans

1860–1931

Austrian astronomer and engineer. In attempting to explain the origin of the Moon, Hoerbiger proposed that the Earth had originally captured four moons consecutively. Three of those moons he said, crashed into the Earth and shattered. Those smashing into Earth are what he believed created the Atlantic, Indian and Pacific oceans. The last moon is the one we see today. Hoerbiger believed it to have originally been a small planet that travelled too close to the Earth and was caught by Earth's gravity. He held that when this last moon was captured, it became Earth's **satellite** and in the process caused chaos on the Earth. Hoerbiger maintained that the lunar **surface** is overlaid with a dense coating of ice. He believed that when he was observing the Moon, he was looking at ice on the surface, and thought it to be several miles deep. Sometime around 1913, he wrote a book with a schoolteacher friend by the name of Philip Fauth titled *Doctrine of Eternal Ice*, in which he wrote of his moon theory. His hypotheses were later further developed and expounded upon by researcher and author **Hans Schindler Bellamy**. Eventually Hoerbiger's ideas were disproven.

Hollow Moon Theory

In November 1969, during the ***Apollo 12*** mission, the **astronauts** set up seismometers on the lunar **surface**. After the **astronauts** returned to the Command Module, they deliberately crashed the Lunar Module into the Moon. The impact with the Moon was of a strength comparable to one ton of TNT. The shock waves lasted for nearly sixty minutes. NASA scientists were quoted as saying that the Moon "rang like a bell." The Moon reverberating for so long was surprising and could not be explained. Maurice Ewing, one of the directors of the seismic testing, at a press conference stated, "As for the meaning of it, I'd rather not make an interpretation right now, but it is as though someone had struck a bell, say in the belfry of a church, a single blow and found that the reverberation from it continued for 30 minutes." There was speculation afterward that the Moon was hollow. The idea of the Moon being hollow sparked the imagination of many and confounded others, because if the Moon were hollow, then it could not be a natural **satellite**. Thoughts and speculations were brought to light as to what the Moon might be and questions pertaining to its origin were discussed.

Some of the theories centering on the possibility of the Moon being hollow included it being a hollowed out **planetoid,** with a civilization is living inside. In a similar line of thought, there is a hypothesis that the Moon is a spaceship, with crew members living and working inside of the Moon's interior, complete with all the facilities and utilities required to run a ship for enormous amounts of time. Russian researchers Mikhail Vasin and Alexander Shcherbakov popularized the spaceship idea, proposing that the Moon was once a planetoid that was hollowed out with gigantic equipment that was used by **extraterrestrials** that were more technically advanced than humans. The famed astronomer and cosmologist Carl Sagan, in his 1966 work *Intelligent Life in the Universe,* stated, "A natural **satellite** cannot be a hollow object." In author Jim Marrs' book *Above Top Secret* (page 226), NASA scientist Dr. Gordon MacDonald is quoted from 1962 stating, "If the astronomical data are reduced, it is found that the data require that the interior of the Moon be less dense than the outer parts. Indeed, it would seem that the Moon is more like a hollow than a homogeneous sphere." See **Farouk El Baz; Spaceship Moon Theory.**

Hologram Conspiracy Theory

The theory that the Moon is either a hologram or is overlaid with a hologram in order to keep its true nature a secret.

Horizon

The horizon as seen on the Moon is closer than what we witness on the Earth. The Moon, being a smaller sphere than the Earth, has more of a curved **surface**, which means a closer horizon. In fact, on the Moon, one would see the horizon at about half the distance as one would on the Earth.

Horns (of the Moon)

(Also Horns of the Crescent)

The points on each side of a **crescent moon** are sometimes referred to as the "Horns of the Moon." While the crescent Moon and its horns represent light and happiness, there is also a darker representation. The name Lucifer (the Biblical Satan and fallen angel) is a negative form of old tales where the Moon is represented as "the "Horned One." Moon watchers through the centuries have observed mysterious lights between the horns of the crescent. They are sometimes referred to as "lights" and also as "stars." There is no explanation for what they may be and they remain a mystery to this day. See **Cotton Mather.**

Hubble Telescope Conspiracy Theory

The Hubble Telescope (named after astronomer Edwin Hubble) is a large space telescope that was launched into a low Earth orbit by NASA in 1990. It travels around the Earth at approximately five miles per second, taking photographs that reach far into the cosmos. It has observed galaxies, planets and stars that are trillions of miles away from Earth. It has shown stars being born and stars that are dying. There are those who would like to see pictures of the man-made objects that are left behind on the Moon (e.g., the base of the Lunar Module–the Eagle). It is believed that the Hubble Telescope could easily provide those images. Requests for these pictures have not been fulfilled. The fact that no photographs have been provided of items on the Moon from the Apollo landings has fueled the argument of those who believe that the Moon landings were fabricated. The official reason given for the failure to provide the requested pictures from Hubble is that, even though Hubble is a powerful telescope, it is not able to differentiate between the **Apollo** gear and the surrounding natural formations of the Moon.

Human Body

The human body is 60 percent water. Some believe that since the Moon has an influence on the oceans' tides, that it also has an effect on the human body and the brain. A **full moon** is thought to affect emotions, causing heightened feelings, passions and sometimes negative thoughts. It is a time believed by some to be when a person is most likely to carry out a crazy and extreme deed. While this view was accepted in years past, scientists today often deny it to be true. See **Lunatic**.

Human Colony

See **Alaje of the Pleiades; George Adamski; Alex Collier; Breakaway Civilization; Moon Base; Ingo Swann**.

Humboldt (Crater)

A prominent impact **crater** where according to lunar researchers there appears to be an excavation operation. Reportedly, this operation can be seen on a NASA photograph. The number of said photograph is ***Apollo 15*** 1512640.

I

Illusion

See **Moon Illusion**.

India

See ***Chandrayaan-1.***

Indian Space Research Organisation (ISRO)

See ***Chandrayaan-1.***

Industrial Complex

A group of anomalous objects found in a NASA photograph. The photograph has formations that appear to resemble structures and roads with lights. Images can be found at *The Living Moon*: http://www.thelivingmoon.com/43ancients/02files/Moon_Images_Book02.html.

Irwin, James Benson

1930–1991

Former **Apollo** astronaut and the eighth man to walk on the Moon. Irwin also served in the United States Air Force as a pilot; he was an aeronautical engineer as well as a test pilot. He was the Lunar Module pilot for ***Apollo 15***. During his visit to the Moon, while working on the lunar **surface** with astronaut **David Scott**, he witnessed a strange object streaking by that scarcely missed hitting them. What it was that nearly struck the two men is still a mystery.

"Is the Moon a Creation of Alien Intelligence?"

A paper written by Soviet scientists, Alexander Shcherbakov and Michael Vasin. Both men belonged to the distinguished Soviet Academy of Sciences. The two published their hypothesis in July 1970 in an article for the Soviet magazine *Sputnik*. It was titled "Is the

Moon a Creation of Alien Intelligence?" They stated that the Moon is a **planetoid** that was hollowed out in the ancient past, somewhere outside of our solar system. They maintained that extraterrestrials used large equipment to dissolve stone and create enormous chasms inside the Moon. Magma from the melted rock poured onto the lunar floor. The resulting massive ship was navigated across the universe and ultimately placed within Earth's orbit. Shcherbakov and Vasin said, "It is more likely that what we have here is a very ancient spaceship, the interior of which was filled with fuel for the engines, materials and appliances for repair work, navigation instruments, observation equipment and all manner of machinery…in other words, everything necessary to enable this 'caravelle of the Universe' to serve as a **Noah's Ark** of intelligence, perhaps even as a home of a whole civilization envisaging a prolonged existence and long wanderings through space."

Islam

See **Black Line Phenomena**; **Splitting of the Moon**.

J

Jade Rabbit

(The Yutu)

An unmanned Chinese lunar probe. The Jade Rabbit was launched on December 1, 2013. Reportedly, there was a problem with the Jade Rabbit and it ceased functioning. In his book *The Moon is Hollow and Aliens Rule the Sky* (page 5), author Rob Shelsky states, "We should note here that even the most recent unmanned robotic landing, in December 2013, the 'Jade Rabbit,' and one made by the Chinese, has quickly ceased to function correctly. It has so far failed to even come close to completing its planned mission as a result." The reason for the Jade Rabbit ceasing to function is unknown, although theories were put forward. According to sources such as *Yahoo News UK*, many of the Chinese, including experts in the field, attribute the failure of the Jade Rabbit on **UFO** interference. Scientists working on the project maintain that a UFO may have been the culprit behind the spacecraft no longer working, in order to protect the secret of there being **bases** on the Moon. Officially, the reason behind its failure is mechanical problems.

Jastrow, Robert

See **Rosetta Stone of Planets**.

Job

See ***Bible***.

Johnston, Ken

1942–

Former NASA consultant. During the **Apollo program**, Johnston supervised the Data and Photo Control Department at NASA in Houston, Texas. Johnston came forward with a number of reports of mysterious occurrences associated with the space program. According to Johnston, the **astronauts** photographed unusual objects that they had discovered on the Moon. Johnston is also said to have viewed an ***Apollo 14*** film which showed a **crater** in which there were five illuminated **domes**. Purportedly, one of the domes had a substance emanating from the top of it that some believe may be steam.

Jyotsna

An Indian goddess found in Hindu beliefs. She is the goddess of the autumn moons and is closely associated to the twilight of the day.

K

Kareeta

An object referred to as a "mechanical bird" that was seen flying across the Moon on October 9, 1946 by residents of San Diego, California during a meteor shower. The name "Kareeta" was telepathically given to Mark Probert (a psychic) from benevolent beings aboard the craft. Mark Probert along with eleven other people witnessed the strange object. An article was written about the Kareeta in the *Round Robin* journal in 1946. See **Layne Meade**.

Kepler (Crater)

A Copernican-aged lunar impact crater, flanked by **Oceanus Procellarum** and Mare Insularum. It was named after famed German astronomer Johannes Kepler (1571–1630). Kepler made pertinent observations about the Moon including the Moon's **orbit**, as well as the characteristics of the lunar **surface**. Purportedly, the Kepler crater has an odd unnatural formation near it that is in the shape of a **cross**. It measures four miles long and half a mile high. Some wonder if the strange construction is a fluke of nature, or if it has a higher meaning, placed there by advanced intelligent beings long ago. Researchers believe it to be an optical illusion, appearing in a cross-like shape due the combination of lighting and shadows. In the book *Elder Gods of Antiquity* (page 102), author M. Don Schorn writes, "author Peter Kolosimo reported that photographs of strange structures in the shape of a cross, were taken by astronomer R.E. Curtis and published in the *Harvard University Review*. They were deemed to be too purposefully arranged to be merely natural formations." Additionally, on February 5, 1884 a mysterious illumination was seen in Kepler. The origin of the light remains unknown.

Kepler, Johannes

See **Kepler Crater**.

Keyhoe, Donald Edward

1897–1988

Prominent writer in the field of ufology. He was a former American Marine Corps Naval Aviator and the former head of the National Investigation Committee on Aerial Phenomena (NICAP). He is the author of several books including *The Flying Saucers are Real* (1950), *Flying Saucers from Outer Space* (1953), *The Flying Saucer Conspiracy* (1955), *Flying Saucers: Top Secret* (1960) and *Aliens from Space: The Real Story of Unidentified Flying Objects* (1973). Keyhoe once talked about ***Gemini 1*** (the first of the Gemini missions). He relayed that *Gemini 1* was followed by four **UFOs** as it headed towards orbit. Keyhoe obtained this news from scientists who were working at the time at Cape Kennedy and were monitoring *Gemini 1* on radar during this period. The scientists told Keyhoe that the UFOs followed *Gemini 1* and stayed with it for a full revolution. Eventually, they abruptly flew off and vanished. The Gemini UFO incident was to be the first of many sightings associated with the NASA missions and a prelude to future Moon missions and their experiences with UFOs. In the book *Anunnaki Theory Authenticated with a New Twist: Historic Evidence of Anunnaki Presence,* by Shafak GokTurk, Keyhoe is quoted as asking, "All clues point out to not only the existence of a Moon Base, but also the operations of a smart race, which have already begun. If this is true, who could these entities be? Do they come from other planets, or do they originate on the Moon?"

Khonsu
(Also Chons, Chonsu, Khensu and Khonshu)

A god of the Moon in upper ancient Egypt. His name means "wanderer." He is the son of the god and goddess Amun and Mut. Khonsu wears a crown of a **crescent moon** cradling the orb of a **full moon**.

Klehanoai

The god of the Moon in Navaho beliefs. The name Klehanoai means "night bearer." He was created in the beginning, along with the dawn, from a beautiful and powerful crystal. His face is covered with a brilliant sheet of lightning. The orb of the moon shields him as thetravels unseen across the darkness of night.

Knowth, Ireland
See **Lunar Map at Knowth**.

Koran
See **Black Line Phenomena**; **Splitting of the Moon**.

Kozyrev, Nikolai Alexandrovich
1908–1983

A leading Russian astronomer and astrophysicist. He is listed as being with the Pulkovo Observatory and also the Crimean Astrophysical Observatory. On November 3, 1958, Kozyrev photographed a mysterious spectrum of a reddish patch of light on the Moon. He stated that the light traveled and then disappeared after an hour. He also reported on an earlier occasion that he had seen what looked like a luminous "cloud" near the central peak of the **Alphonsus crater**. It was thought by some to be caused by volcanic activity in the area, or even gas spewing forth from the **crater**'s central peak. While that report cannot be discounted, nor the volcanic activity theory proven, **UFO** researchers suggest that a light that moved and then vanished is highly irregular. A second account states that Kozyrev reported seeing gas in 1958 and 1961 in the vicinity of the Aristarchus crater where spectrograms revealed the existence of carbon and hydrogen gas.

Kraft Jr., Christopher Columbus
1924–

Former NASA engineer and manager. He directed the NASA tracking base in Houston, Texas throughout the **Apollo** missions.

Reportedly, Kraft released tapes where the ***Apollo 11*** astronauts were speaking about seeing **UFOs** during the mission.

Kubrick, Stanley
1928–1999

American screenwriter, director and producer. Examples of Kubrick's work include the classic movies ***2001: A Space Odyssey***, *The Shining* and *Eyes Wide Shut*. There is a story that Kubrick once stated in a videotaped interview, that he worked with NASA in faking the ***Apollo 11*** moon landing. A video of the interview was released fifteen years after Kubrick had passed away. Writer and director T. Patrick Murray is said to have interviewed Kubrick on tape a few days before he passed. The interview is believed to be a hoax.

L

Laplace, Pierre-Simon

A noted physicist and astronomer from France. He is renowned for his analysis of the stability of the solar system. It was Laplace that put forward the **Coaccretion Theory**, also known as common birth theory and the condensation hypothesis. It is the idea that the Earth and Moon were created at the same time, independently from each other, from a nebular cloud of dust and gas that coalesced over time. Laplace considered the Earth and Moon to be a binary system that revolves around the Sun jointly.

Last Person on the Moon

Astronaut **Eugene Cernan**, who served as Commander of ***Apollo 17***, was the last man to walk on the Moon in December of 1972.

Layers

The layers of the Moon are arranged in a very mysterious way. Typically the bottom layers of an astral body are denser than the top (surface) layers. On the Moon, the top three layers are reversed, the densest layer being on the **surface**.

Layne, Newton Hathaway Meade
1882–1961

Publisher of the *Round Robin* journal, parapsychologist and **UFO** researcher. In October 1946, in the *Round Robin*, Layne writes about an unusual event involving the Moon. It was on October 9, 1946, on an evening in San Diego, California when there was a meteor shower. Layne wrote of receiving a call from Mark Probert, who, with a friend, was watching the meteor shower from the top floor of a building when he spotted an unusual flying object in the sky. Probert described the strange object as being as large as a plane, with "two reddish lights." The second time he saw it, the object moved across the face of the Moon. He stated that it was "luminous" and oddly had "flapping wings." Probert's friend, Fernando Estevane, stated also that the object had two lights and was not a plane. Estevane also commented that a regular airplane would not be able to "maneuver" the way that this had. The following day, on October 10, eyewitness accounts were reported to the *Round Robin*. Mark Probert reportedly was a psychic and was able to channel a message from the ship, through clairaudience. Through his psychic ability, he was told that the ship came from "west of the Moon," that the occupants "come in peace" and they were more advanced than the people of Earth. The craft was identified as a "mechanical bird called the Kareeta." The psychic reported that the craft came from a distant planet and the occupants were reluctant to land because they were not certain of the reception they would receive on Earth.

Lemuria

(Mu)

An alleged lost continent that existed in the Pacific Ocean, said to have been sunk by one of ancient Earth's two moons. According to tales of old, there was a time when the Earth had two moons. It was one of those moons that is said to have destroyed the ancient kingdom of Lemuria. Lemuria is believed to have existed side by side with **Atlantis**. The two competed for prominence and to be the superior nation. The Atlanteans thought it necessary to gain the advantage by destroying Lemuria, also known as Mu. During this same period, advanced beings from another world had an association with mankind. Some posed as deities, others as friends. From where these beings came has been lost. The Atlanteans enlisted the aid of these **extraterrestrials** in helping to destroy Lemuria. They chose a strategy that involved the Earth's moons. Through superior technology, they captured one of the two moons. The Atlanteans and the extraterrestrials moved the moon from orbit with force fields, putting it as close to the position of Lemuria as possible and then destroyed the moon. This destruction caused meteorites to rain down on Lemuria destroying most of it, along with its people. In the upheaval and chaos, there was also an explosion from beneath the Earth, causing Lemuria to sink into the ocean, never to be seen again.

Leonard, George H.

1921–

Former NASA scientist and photo analyst, who once worked in the photographic intelligence division of NASA. He is the author of *Alien*, *Alien Quest* and *Somebody Else is on the Moon*. While working at NASA, Leonard was assigned the job of examining Moon photographs taken by the space probes that NASA sent up in the sixties. NASA had been searching for safe landing areas for future **astronauts**, as well as mapping the Moon. In the process of his studying the photographs, Leonard came across pictures that looked to him to contain several unexplained oddities on the lunar **surface**. In his book *Somebody Else is on the Moon,* Leonard published the result of his in-depth investigation of thousands of photographs of the Moon in which he found mysterious objects and symbols. Leonard claims to have found large mechanical rigs in operation on the lunar floor, odd geometric shapes and symbols, tall constructions and towers, track marks running along the lunar floor, huge **domes** in the middle of what are thought to be unnaturally lighted **craters** and more. It is his belief that these photographs show proof of what he believes to be the result of **extraterrestrials** visiting the Moon. In *Moongate: Suppressed Findings of the US Space program* by William L. Brian II (page 147), it states: "George Leonard supplied photographic evidence that the Moon is being worked with massive machines in his book *Somebody Else Is on the Moon*. He suggested that damage done to the Moon's surface at one time is being slowly repaired." In *Somebody Else Is on the Moon* (page 20), Leonard writes, "The Moon is firmly in the possession of these occupants. Evidence of Their presence is everywhere, on the surface, on the nearside and the hidden side, in the craters, on the maria and in the highlands. They are changing its face." Skeptics contend that the objects in the pictures that Leonard studied were the combined effects of light and shadows.

Lightening

On May 18, 1787, two astronomers reported seeing lightening above the lunar **surface**. That event caused some to speculate that the Moon had an **atmosphere**. The cause of the mysterious lightening remains unknown.

Linne (Crater)

A small, young impact **crater**. German

astronomer **Johann Schroeter** monitored and chronicled unexplainable changes in the six-mile, 1200 feet deep crater. Over the years, as he documented his observations, he watched Linne slowly diminish. Linne is now much smaller with very little depth. The mystery of what happened to Linne remains. NASA photographs taken by ***Apollo 15*** place Linne at 1½ miles across. Schroeter, who believed that the Moon was inhabited, attributed the shrinking of Linne to work being performed on the Moon by **Selenites**.

Lippert, Rudolph M.
(Unknown birth and death dates)

On September 16, 1953 astronomer Rudolph M. Lippert, a member of the Lunar Section of the British Astronomical Association, saw a brilliant flash hit the Moon. Afterward, he saw a light that glowed yellow-orange. The mysterious event is believed by Lippert to have possibly been a meteorite striking the Moon's **surface**.

Litter
See **Man-made Materials**.

Littrow (Crater)

An impact **crater** named after Czech-born Joseph Johann von Littrow. Littrow was a director of the Vienna Observatory from 1819 to 1840. The Littrow crater is located in the northeastern area of the Moon's nearside, on the eastern rim of Mare Serenitatis. On January 31, 1915, seven bright spots in a formation resembling the Greek letter gamma was observed inside the crater. On September 13, 1959, astronomers sought to photograph the Littrow crater and vicinity. They were unable to do so because of a mysterious large, dark, unexplainable, cloud-like mass that was obscuring the crater.

Littrow, Joseph Johann von
1781–1840

Austrian astronomer. He was once the director of the Vienna Observatory. It is said that Littrow sought to alert possible Moon inhabitants to our presence and communicate with them. His method included digging trenches in geometric shapes in the Sahara, filling them with water, topping them with kerosene and then lighting them. The Moon's **Littrow crater** is named in his honor.

Lobachevsky (Crater)

An impact **crater** on the **dark side** of the Moon. Per an independent study, there are a number of anomalies found in old NASA photographs that stand out in the Lobachevsky crater, as imaged by the ***Apollo 16*** mission. Steve Wingate, an American researcher, is credited with discovering the strange abnormalities and bringing them to the public's attention. There are said to be several unusual objects and constructions at Lobachevsky. One is a strange light-colored, oblong construction that runs along the edge the crater rim. Another is a huge obelisk that stands hundreds of meters tall. There is also an unexplainable configuration sitting on the floor of what looks to be a canyon located beneath the crater's edge. In addition, there is an odd formation that appears to have a luminous, mirror-like **surface**. Various theories have been applied as to what these all are. The hypothesis include residue from volcanic activity, photographic irregularities, a mining excavation, debris from an asteroid hitting the Moon, **extraterrestrial** artifacts and even an extraterrestrial **base**.

Locke, Richard
1800–1871

A journalist for *The Sun* newspaper located in New York City during the 1800s. Locke became famous for perpetrating a Moon hoax in a series six of articles, published in *The Sun*, in 1835. Locke was able to convince the public that it had been discovered that the Moon is inhabited. In the articles, he credited famed astronomer of the day John Herschel for making the discovery. See **Great Moon Hoax**.

Longhorn, The

An anomalous construction in the **Tycho crater**, as featured in author **Mike Bara's** book *Ancient Aliens on the Moon* (page 188). The longhorn was located in an image taken by the ***Clementine*** probe. Bara describes the longhorn as "a basically symmetrical object with two central 'nodes' and curved arms extending from the central body. There appears to be some underlying support just to the left of the right hand carved 'arm,' but the central spherical 'node' looks to be above the ground, judging by the shadow beneath it." Bara states, "I am not aware of any accepted process that could account for this object forming naturally." The longhorn is one of many of alleged pieces of equipment that lunar researchers have located in images of the Moon. Some suspect that there is excavation going on, carried out by unknown beings.

Lost Cosmonauts

See **Cosmonauts**.

Lovell Jr., James Arthur

1928–

Former NASA **astronaut.** Additionally, Lovell served in the United States Navy as a captain and was a Naval Aviator. He served on ***Gemini 7***, Gemini *12*, ***Apollo 8*** and was commander of the hapless ***Apollo 13*** mission. Lovell saw a mysterious object in space during the *Gemini 7* mission. In December 1965, Lovell and astronaut **Frank Borman**, spotted a **UFO** during the course of their second orbit. Borman reported it and stated that it was quite a distance away from their capsule. During the *Apollo 8* mission, after returning from the **dark side** of the moon, Lovell made a comment that has been a subject of debate. Lovell stated, "Mission Control, please be informed, there is a Santa Claus." This statement was made during Christmastime in 1968 and many feel that it was an innocent comment made in spirit of the holiday. Others maintain that Santa Claus is a code word used by the astronauts for UFOs.

Lucian of Samosata

120 CE–180 CE

Notable Syrian satirist. Lucian of Samosata wrote over eighty works in his lifetime. An educated, intelligent and well-traveled man, Lucian mentioned a civilization in one of his works known as the **Arcadians.** Reportedly, the Arcadians existed before there was a moon. Lucian stated in his writing on Astrology, "The Arcadians affirm in their folly that they are older than the moon." His words lend credence to the argument that there was a time before the Earth had a moon and for some researchers, supports the theory that the moon is not a natural **satellite**.

Lucina

A Roman goddess of light. She is connected to the radiance of the Moon.

Luminosities

A name given to small luminous orbs of light that are often spotted on the Moon and are sometimes referred to as "**bright spots**" or "stars." These brilliantly glowing orbs have been reported over two hundred times. The **Plato crater** is said to be the place where these mysterious lights appear the most often. One report states that there have been dozens of these orbs found in that location, and some refer to it as a "hotspot" of activity. British astronomer **Cecil Maxwell Cade**, who was once a member of the prestigious Royal Astronomical Society and author of the book *Other Worlds Than Ours*, wrote about these mysterious lights: "Star-like lights, which could not have been due to the Sun's rays illuminating the tops of high mountains, have been the subject of many hundreds of observations; in fact, up to April 1871, no fewer than 1600 observations had been made for the **crater** Plato alone." He continued his comments stating that the lights were sometimes seen in small groups and even in geometrical arrangements. He pondered whether beings living on the Moon, or even visiting the Moon, were trying to reach the people of Earth. In the end he states, "Both of these notions are extremely improbable, to say the least, but no one really knows what they were."

Luna (1)

The Latin word for Moon in ancient Rome.

Luna (2)

The name of the ancient goddess of the Moon in Roman Mythology.

Luna (3)

The name given to an alleged extraterrestrial **base** located on the **far side** of the Moon. Purportedly, the Luna base has facilities and a mining operation. It is also an area where **extraterrestrial** spacecraft are believed by some to be parked. The ships are said to include both large motherships and smaller **spaceships** that are used as shuttle craft.

Luna 9

The first spacecraft to perform a soft landing on the Moon. *Luna 9* was an uncrewed space mission from the Soviet Union's Luna Program. It launched on January 31, 1966 and landed February 3, 1966. It was the first space mission to send close-up images of the Moon's **surface**. Reportedly, *Luna 9* took a picture of structures in geometric shapes that appeared to have been placed on the lunar surface for a purpose. In an article from *Argosy Magazine* (August 1970) titled "Mysterious 'Monuments' on the Moon," it states: "Four years ago, Russia's *Luna 9* and America's *Orbiter 2*, both photographed groups of solid structures at two widely separated locations on the lunar **surface**. These two groups of objects are arranged in definite geometric patterns and appear to have been placed there by intelligent beings."

Luna Incognito

An area of the Moon where very few photographs have been taken. The few photographs there are show an object that some believe is a very large triangular spacecraft.

Lunar Anomalies Report

See ***NASA Technical Report TR-277***.

Lunar Cavern(s)

A large chamber located inside the Moon. There are believed to be caverns located within the interior of the Moon. Additionally, there are openings on the **surface** of the Moon that lunar experts believe may lead into caverns. Moon observers maintain that some type of life form could exist inside the caverns, perhaps a civilization within the Moon. Astronomers of old believed such things were possible. Today, scientists approach from a more practical view and maintain that the Moon cannot sustain life. In Jim Marrs' book *Above Top Secret* (page 228), prominent geologist **Farouk El Baz** who was once the secretary of the landing site selection committee for NASA's **Apollo program**, is quoted as saying, "There are many undiscovered caverns suspected to exist beneath the **surface** of the Moon."

Lunar Color Phenomena, ACIC Technical Paper No. 12

In May of 1964, the Lunar and Planetary Branch, the Cartography Division of the United States Air Force Aeronautical Chart and Information Center, published a report titled "Lunar Color Phenomena, ACIC Technical Paper No. 12." The report featured a spectacular red-colored **transient lunar phenomenon** which had materialized near the **Aristarchus crater** on October 29, 1963 and on November 27, 1963. The purpose of the report was to record the two mysterious occurrences, devoid of opinion behind the cause. The creators of the report believed that what had been witnessed was of great scientific value. No conclusions were given as to the reasons behind the strange sightings. A number of hypotheses were given by lunar specialists as to what the phenomena could be. See A**ristarchus Crater; Edward Barr; James Greenacre.**

Lunar Crater Observation and Sensing Satellite (LCROSS)
See **Bombing**.

Lunar Craters
See **Craters**.

Lunar Day

The length of a day on the Moon relative to the Sun is approximately 708 Earth hours or 29.5 Earth days. This is the average period of time that it takes the Sun to return to the same spot in the sky as seen from the surface to the Moon. This period is called synodic day, which should not be confused with the sidereal day.

A sidereal day is the period that it takes the moon to make a full rotation on its axis—about 27.3 Earth days. Because the Moon and the Earth are tidally locked (i.e., the Moon always points the same face towards the Earth), this is the same period of time it takes for the Moon to make a full orbit around the Earth. However, since, during the same period, the Earth-Moon system also rotates around the Sun, it takes an additional 2.2 Earth days for the Sun to return to the same position in the sky of the Moon.

Lunar Eclipse

A lunar eclipse happens when the Moon is obscured as it moves into the Warth's shadow. In ancient times a lunar eclipse was a source of great mystery for some, a cause of fear for others and in some cultures a reason to worship the Moon. From the awe of an eclipse, great tales and myths were born. Today, there are those who still consider the lunar eclipse a mystery. During an eclipse, the Earth is just the right distance from the Moon, to cast a large enough shadow to completely cover the Moon, causing a solar eclipse as seen from the surface of the Moon. Some find it highly coincidental/mysterious that the distances between the Earth and the Moon and between the Earth and the Sun, as well as the diameters of the Moon and the Earth, are relatively such that the Earth would block the Sun completely during a lunar eclipse. Some claim that this is too strange to be a coincidence and contend that it is evidence that the Moon is not a natural **satellite**. Some hold to the theory that the Moon was designed by ultra-intelligent beings and placed in our solar system at just the right coordinates to assist the planet Earth and to support life on it. The popular American author **Isaac Asimov,** who was once a professor of biochemistry at Boston University and who wrote science books for children and adults, once stated, "there is no reason why the moon and the sun should fit so well. It is the sheerest of coincidences and only the Earth is among all the planets blessed in this fashion." Additionally, strange lights have been seen on the Moon during lunar eclipses. See **Appendix 2: Anomalous Light 1800s and 1847; China; Frederick William Herschel; Tetrad; Joseph Zentmayer.**

Lunar Map at Knowth

An ancient map of the Moon that was carved out of stone. Scientists place its age at approximately 5,200 years old. The map was discovered by Dr. Philip Stooke from the University of Western Ontario, Canada. Stooke had been interested in locating a lunar map that was older than the one created by Leonardo da

Vinci in 1505, which at the time was the oldest known map of the Moon. In his quest, Stooke searched through antique writings, history books and documentation from Neolithic sites on the British Isles. He found the ancient map carved into a rock (dubbed Orthostat 47) in a tomb in Knowth, in County Meath, Ireland. In an article written by BBC Online Science Editor Dr. David Whitehouse, Stooke was quoted as saying, "I was amazed when I saw it. Place the markings over a picture of the **full moon** and you will see that they line up. It is without a doubt a map of the Moon, the most ancient one ever found." The map is estimated to be ten times older than the one made by Leonardo da Vinci. Mysteriously, the map offers a thorough diagram of the Moon's characteristics. Stooke hailed the ancient people that created the map as the "first scientists."

Lunar Orbiter 2

A NASA robotic spacecraft. It was a part of the Lunar Orbiter program. Launched on November 6, 1966, its mission was to photograph the lunar **surface**, so that safe areas to land could be chosen for future Surveyor and **Apollo** missions. During its operation, *Lunar Orbiter 2* photographed a cluster of shadows of tall objects on the Moon's surface that were shaped like **monuments** and arranged in geometric shapes. Where these objects came from and what they represent is one of the Moon's greatest mysteries. They have been dubbed the **Blair Cuspids**.

Lunar Orbiter 3

A spacecraft from NASA's Lunar Orbiter program. *Lunar Orbiter 3* was launched on February 5, 1967. Its job was mainly to take pictures of parts of the Moon's **surface** in search of areas where the future Surveyor and **Apollo missions** could land. One of the *Lunar Orbiter 3* photographs revealed a large, erect edifice that resembled a **monument**. This was found in the Moon's Ukert region. It is famously referred to as "The **Shard**." Another curious object photographed by *Lunar Orbiter 3* is known as "the **Tower**." This structure is said to be over a mile thick and approximately five miles high.

Lunar Pareidolia

The phenomenon of seeing a pattern on the Moon when one does not exist. The nearside of the Moon has many dark marks known as maria, which is the Latin word for seas. There is no design to these marks on the Moon. However, when gazing upon the Moon, people often see patterns that resemble figures and objects. Some of the most popular images that people have claimed to have seen include a man, rabbit, toad and a woman. See **Man in the Moon; Rabbit in the Moon.**

Lunar Reconnaissance Orbiter (LRO)

A NASA robotic spacecraft. The *Lunar Reconnaissance Orbiter (LRO)* launched June 18, 2009 and is currently in space. Its mission is to compose high-resolution maps showing the composition of the lunar floor. It is also to look for possible sources of ice and seek out potential safe landing areas for possible future projects involving trips to the Moon. There is a current mystery surrounding the *LRO*. Reportedly, the *LRO* has taken photographs of what some researchers believe may be ancient artifacts from a civilization that was once located on the Moon.

Lunar Transient Phenomenon (LTP)

A title given to anomalous lights on the Moon. It is the same as **transient lunar phenomenon (TLP)**.

Lunarian(s)

A name for Moon inhabitants. See **Selenites.**

Lunate

A term meaning crescent shaped. See

Crescent Moon.

Lunatic/Lunacy

A term used for a psychologically troubled individual. The word "lunatic" is derived from the Latin word "Luna," which means moon. People once believed that an individual who was engaging in rash or psychopathic behavior had been affected by either a **full moon** or the phases of the Moon. For some, this belief persists. There are scientists that maintain that this idea is a myth. The word "moonstruck" also has a similar meaning. When a person is said to be "moonstruck" it implies that the person is exhibiting bizarre, crazy or erratic behavior. Additionally, the term "Loony bin" (an old fashioned term for an institution for the mentally ill) is derived from the Latin word Luna. See **Human Body**.

Lunation

A lunar month (the period of time between new moons).

M

Madler's Square

An anomalous geometric configuration discovered on the Moon near Mare Frigoris. It is named after German astronomer Johann Madler. Because of its design and perfect angles, some believe it to be an artificial structure. Dating back to ancient times, it has extremely high walls and is in an angular-shaped pattern that forms what appears to be a square or rhombus. In *Elder Gods of Antiquity* (page 101), author M. Don Schorn cites a quote by the famed British astronomer Edmund Neison from his book titled *The Moon* (1876). Neison states that Madler's Square is "a perfect square, enclosed by long, straight walls about 65 miles in length and one mile in breath, from 250 to 300 feet in height." This mysterious and strange formation was discussed in many books about the Moon. Some suspect that it was either placed there intentionally by some otherworldly beings, or is the ruin from an ancient civilization from eons ago.

Magnetism

Rock samples taken from the Moon by the **astronauts** were found to be magnetized. This is mysterious since the Moon has no magnetic field. Where the magnetism in the rock samples came from cannot be explained.

Mah

Moon god in ancient Persian mythology. He is often shown positioned in front of a **crescent moon**.

Maimonides

See **Adam**.

Mama Quilla

Incan Moon goddess. Mama Quilla was associated with lunar eclipses.

Man in the Moon

A lunar pareidolia image seen on the nearside of the Moon that appears to some as the face of a person. It has been dubbed the man in the moon and is the most well known of all of the lunar pareidolia in the West. Another lunar Pareidolia image of the man in the Moon includes seeing the entire body of a man. Some maintain that they see the man in the moon carrying thorns. Most interestingly, William Shakespeare wrote of the man in the moon carrying thorns in his play titled *A Midsummer Night's Dream*. Shakespeare writes, "I, the man in the moon; this thorn-bush, my thorn-bush."

An old print of the Man in the Moon.

Mani

The god of the Moon in Norse mythology. He is the brother of Sol, the goddess of the Sun. Mani moves the moon across the darkened night sky, illuminating the way.

Man-Made Materials

Due to humans visiting the Moon, there is plenty of trash, debris and parts of machinery to be found there. It has been predicted that there are around 400,000 lbs. of discarded items and man-made materials on the lunar **surface**. Most of it consists of fragments and rubble from the tests and scientific experiments performed on the Moon. There are small items such as a golden olive branch, a flag kit, golf balls, and the hammer and falcon feather that were used in an experiment that was performed in 1971. There are also larger more substantial objects such as lunar orbiters, robotic probes, five moon rangers, *LCROSS* (the *Lunar Crater Observation and Sensing Satellite*) as well as lunar probes Ebb and Flow from NASA. These materials were left by the **Apollo** missions as well as unmanned missions from agencies from around the world, including the United States, Europe, **India**, Japan and Russia. NASA has a comprehensive list of man-made materials that are on the Moon. It can be accessed online at: history.nasa.gov.

Mapping/Maps

See **British Selenographers; Cartographer(s); Lunar Map at Knowth; Selenographers**.

Marama

A Moon goddess in Polynesian (Maori) myths. According to lore, Marama dies with each moon cycle. However, she is resurrected when she bathes in the waters from which all life originates.

Mare

The singular form of **maria**.

Mare Crisium

(Sea of Crises)

A lunar mare located on the nearside of the Moon in the Crisium basin. Its diameter is 345 miles, with an area of 68,000 square miles. It is approximately 3.9 billion years old. Mare Crisium became the center of attention when what looked to be an artificial **bridge** suddenly appeared there in 1953. Astronomers were certain that the structure had not been there previously. It was first reported by science editor **John O'Neil** of the *New York Herald Tribune*. The bridge was reportedly twelve miles long. Astronomers were confused by the sudden appearance of such a large structure that had not been there previously. British astronomer **Patrick Moore** asserted that the construction had appeared suddenly out of nowhere. The bridge eventually vanished, just as suddenly as it had appeared.

In 1869, the Royal Astronomical Society of Great Britain led an investigation into **anomalous lights** on the Moon. It was reported that strange lights were often seen in and around the **Mare Crisium** region. The lights were witnessed moving, and were often arranged in a variety of patterns including triangular and linear. Some appeared brighter at times than others. There was no explanation as to the source of the lights. Additionally, there have been other mysterious sightings of lights in and around the mare. A compilation follows.

1672, February 3, the mare took on a mysterious nebulous appearance.

1774, July 25, four glowing orbs of light were observed.

1788, September 26, a small nebulous area was seen.

1826, April 12 and 13, a black haze moving over the mare was witnessed.

1832, July 4, small bright points of light and also streaks of light were observed.

1864, May 15, a bright cloud was seen.

1865, April 10, a starlike light appeared. In addition, the mare was covered with streaks of light that intermingled with bright orbs.

1865, April 10, starlike points of lights were seen.

1865, September 5, a number of small lights appeared.

1915, December 11, a small luminous light was observed.

See **Royal Astronomical Society of Great Britain; Bridge; Greaves Crater**.

Mare Serenitatis

(Sea of Serenity)

A lunar mare in the northern hemisphere of the Moon that has been referred to as "spectacular." It is located between Mare Imbrium (Sea of Rains) and **Mare Tranquillitatis** (Sea of Tranquility). It is 416 miles in diameter. It forms the left eye of the "**Man in the Moon**" due to its location and spherical shape. Its base is mainly made up of basalt. Besides its striking beauty, the Sea of Serenity has recently gained attention due to an odd object found in its location. The item allegedly has no natural characteristics and 90 degree angles are seen within its shape. To many observers, the object resembles either a type of structure or spacecraft.

Mare Tranquillitatis

(Also Sea of Tranquility)

A lunar mare that is situated in the Tranquillitatis Basin. It was the landing site of ***Apollo 11*** where **Neil Armstrong** and **Buzz Aldrin** first walked on the Moon on July 21, 1969. The Sea of Tranquility is shrouded in mystery. Reportedly, a strange dark cloud, **spaceships** and odd constructions in the shape of cuspids have all been seen near the sea. On September 11, 1967, Canadian astronomers witnessed a dark cloud with violet edges moving slowly across the Sea of Tranquility. In 1976, **Otto Binder**, a former NASA employee, reported that he had heard one of the **astronauts** informing mission control that there were large alien spacecraft parked in the center of the Sea of Tranquility, where the astronauts were located. In 1978, former NASA employee Maurice Chatelain wrote that there were two **UFOs** hovering above the Sea of Tranquility not long before Neil Armstrong stepped onto the Moon. Chatelain quoted the transmissions of the **astronauts** with mission control. Per Chatelain, the astronauts allegedly told mission control that they saw **extraterrestrial** spacecraft lined up on the remote side of a neighboring **crater**, watching them. It was also near the Sea of Tranquility where six formations dubbed the **Blair Cuspids** were captured in photographs taken by NASAs ***Lunar Orbiter 2***.

Maria

(Singular form is mare)

The substantial, gray-black basaltic areas on the lunar **surface**. The term maria is Latin for seas. Although they have never held water, centuries ago, astronomers believed these large dark areas to be seas, thus the name maria. They are in fact, huge lava flows that congealed into basalt. The existence of the maria is a mystery because these areas point to an enormous torrent of lava in the Moon's ancient past. However, there is no evidence that the Moon could ever have had any volcanic activities—was never hot enough.

Maskelyne, Nevil
1732–1811

British Astronomer. Maskelyne is known for being the first to scientifically gauge the Earth's weight. In 1794, Maskelyne witnessed a mysterious cluster of lights traveling across the dark half of the Moon. His sighting was reported to the Royal Society.

Mass

The mass of the Moon is $7.342 x 10^{19}$ Kg, about 81 times lighter than the Earth's.

Mather, Cotton
1663–1728

In November of 1668, a light was witnessed on the moon by Minister Cotton Mather. In a letter to the Royal Society in Boston dated November 24, 1712, he stated, "ye star below ye body of ye Moon and within the Horns of it... seen in New England in the Month of November, 1668." The letter is now located in the library of the Massachusetts Historic Society in Boston.

Mattingly, Thomas Kenneth (Ken)
1936–

Former NASA **astronaut**, serving as the Command Module pilot for ***Apollo 16.*** In addition, Mattingly was a Rear Admiral for the United States Navy, an aeronautical engineer and test pilot. During the *Apollo 16* mission, Mattingly was startled to see a **UFO**. It disappeared behind the Moon's horizon as he was orbiting the Moon.

Mayans

An ancient Mesoamerican civilization. The Mayans are well-known for their arithmetic, art, architecture, astronomical system, calendar and hieroglyphic writing. The Mayans left records that talked about a time when there was no Moon. They wrote that it was Venus that shined in the night sky.

McDivitt, James Alton
1929–

Former NASA **astronaut.** McDivitt was the Command Pilot for ***Gemini 4*** and the Commander for ***Apollo 9***. He was also a Brigadier General, American test pilot, aeronautical engineer and US Air Force pilot. In June 1965, McDivitt and astronaut **Edward White**, observed a luminous, **cylindrical-shaped** object while in space. At the time, they were in the *Gemini 4* capsule. The object was in close proximity to their position. It was so close that the **astronauts** worried about a possible collision. They reportedly went so far as to plan to take evasive action if it came any closer to their spacecraft. This sighting occurred as they travelled over Hawaii. McDivitt was able to capture the **UFO** on film as it passed by. The object was described as a glimmering, silver-colored, cylindrical UFO with elongated appendages stretching out from it, heading towards the craft. In an interview from the 1960s on *SpaceTimeForum*, McDivitt talked about the sighting. He was asked if he had ever seen a UFO. He stated, "Yes, during my flight on *Gemini 4* in 1965, I saw an object in space fairly close to our spacecraft, that I could not identify." Years later, McDivitt, in an interview with the NASA *Johnson Space Center for their Oral History Project*. On June 29, 1999 he stated, "I had no idea whether it was a little thing up close to the window or if it was a big thing out a little bit further. It could've been the size of the Empire State Building for all I knew way out there. But I'm sure it was in the—in our orbit and it probably was a piece of ice that had fallen off the spacecraft someplace. Or maybe a piece of Mylar."

Meier, Billy

1937–

An extraterrestrial **contactee** of the Pleiadians. Meier claimed to have communicated at great length with **extraterrestrials**. He photographed numerous **UFOs**, which he used as evidence that his experiences were real. He recorded information that was given to him by a humanoid female from the Pleiades named Semjase, said to be 344 years old. Meier described her as young and slender, with fair skin, blue eyes and blond hair. Her only distinguishing feature that showed she was not human, was the shape of her ears, which were described as longer than a humans. According to the information that Meier was given, the Moon originated as a small planet and was brought into our solar system by a comet.

Men

A Moon god of the Phrygian people, Western Asia Minor. He appears in art with the horns of the **crescent moon** behind his shoulders.

Menger, Howard

1922–2009

An extraterrestrial contactee who claimed to have travelled to the Moon in a spaceship. Menger allegedly had contact with otherworldly beings on and off from the age of ten. He was a well-liked, charismatic speaker that people listened to as he recounted his experiences with extraterrestrials. He likened these space travelers to our own **astronauts**, stating that they were simply passing through, just as our astronauts were when visiting the Moon. His experiences led to his writing such books as *From Outer Space to You* and *The High Bridge Incident*. Menger once told the story of his meeting in 1956 with extraterrestrials. He recounted that he had carried a Polaroid camera along with him and photographed the extraterrestrials and their spaceship. Afterward, he was taken on a trip to the Moon. Menger commented in his book *From Outer Space to You* saying, "As my space friends had promised, they took me on my first trip to the moon the second week of August, 1956." Menger took pictures of the Moon through the window of the spaceship, from which he saw **structures** on the Moon. On his second trip to the Moon, Menger took pictures while approaching the Moon and while on the **surface**. His pictures can be seen in, *From Outer Space to You*. Menger and others from Earth, were taken for a tour aboard a vehicle resembling a high tech locomotive. It was wheelless, each car was covered with a **dome** and the roadway was made of copper. The vehicle was suspended above the ground as they glided silently and smoothly along. Menger was also shown underground facilities, and natural scenery such as mountains and valleys. There was one area that Menger said reminded him of the Valley of Fire in Nevada (the Valley of Fire takes its name from red sandstone formations). He described the color of the sky as saffron combined with yellow and the lunar floor as a pale yellowish color. The surface material he described as powdery. There were also rocks and boulders to be found and small patches of vegetation that were scattered intermittently. Menger and the other guests were taken to various buildings which were all domed. They witnessed areas of art, architecture and more. At one point on the tour, the escort allowed the guests to experience the outside nature of the Moon. They momentarily put their heads outside and experienced the extremely hot temperatures on the Moon. Menger recalled thinking that no one would be able to live outside on the **surface** of the Moon for long.

Mercury 9

The last of the Project Mercury missions. The Mercury missions were a prelude to the future **Apollo** missions that would put men on the Moon. It launched on May 15, 1963. A mysterious voice in an unknown language was picked up on a special radio frequency by astronaut **Gordon Cooper** on May 16, while passing over Hawaii. Later, the recording was examined and it was determined that no known earthly language matched what was on the tape. In addition, as he crossed over Australia, Cooper saw a huge green **UFO** that was also picked up by tracking stations. No official word came out on what the UFO was. UFO proponents believe that the space program is being watched by beings from another world, and that they are especially interested in man's progression into space, particularly their trips to the Moon. It is thought that what Cooper picked up on the radio, as well as the UFO, may be explained by that hypothesis.

Mercury-Atlas 6
See **Glenn Jr., John Herschel**.

Military Base
See **John Brandenburg**.

Mining

The idea that the Moon is being mined stems back to a story that centered around the ***Apollo 11*** astronauts allegedly seeing mining equipment on the Moon. Some moon researchers have attempted to tie the Moon being mined into its ancient history. There are those that hold that there were **extraterrestrial** traders in the ancient past that mined the Moon and traded their findings with other beings. Extraterrestrial contactee **George Adamski** mentioned that during his trip to the Moon, he witnessed a barter system where supplies were traded for minerals from the Moon with others that flew in on **spaceships**. The assumption is that the Moon was excavated to obtain these minerals. There is also a theory that extraterrestrials built **bases** on the Moon to reside in while mining the Earth. According to researcher and writer Don Wilson in his book *Our Mysterious Spaceship Moon*, what looked to be excavating operations were located on the Moon by the Apollo **astronauts**. He writes, "The concentric hexahedral excavations and the tunnel entry at the terrace side can't be results of natural geological processes; instead, they look very much like open cast mines." Open cast mining is a **surface** mining technique in which large pieces of land are hollowed out to allow for extracting minerals. Some believe that **Helium 3** may be the source of interest. See ***Apollo 14***.

Missions
See **Astronauts; Apollo Missions**.

Mita, Koichi
1885–1943

A Japanese psychic who specialized in Nensha (psychic photography, where the person has the ability to burn pictures onto a surface with thought). It is also known as thoughtography, thermograpy and nengraphy. Koichi Mita became adept at the skill in the early 1900s. He first learned of Nensha in 1914 and quickly perfected his newfound gift. By 1916, he was displaying his ability in front of large audiences, burning such recognizable images, as the Ogaki Castle and Japanese kanji characters onto photographic plates by thought. Mita, wanting to take his abilities even further and prove to skeptics that he was wholly legitimate, choose to project images of the **dark side** of the Moon. On June 24, 1931, he was successful in his pursuit of excellence and created images of the Moon's dark side, burning them onto two plates. The first photographs of the dark side of the Moon came in 1959, taken by the Soviet's *Luna 3* space probe. Dr. Goto Motoki who was the president of Japan's Agency of Industrial Science and Technology from 1960 to 1961, studied Mita's images and confirmed that they were completely accurate.

Mitchell, Edgar Dean
1930–2016

Former NASA **astronaut**. Mitchell served on ***Apollo 14*** as the Lunar Module pilot. He was the sixth man to walk on the Moon, exploring the Fra Mauro Highlands area for nine hours. Before becoming an astronaut, Mitchell was a United States Navy officer and aeronautical engineer. In 1970 he was the recipient of the Presidential Medal of Freedom for an attempted mission to the Moon. In his later years, Mitchell was a **UFO** researcher and author, and founded the Institute of Noetic Sciences. Mitchell participated in numerous interviews and several documentaries, speaking on his ideas about the existence of **extraterrestrials** and UFOs. In his book *Earthrise* he tells of his mission to the Moon and explains how traveling there gave him a different outlook on life and an understanding that we are not alone in the universe. Mitchell once recalled an experience that he had while working on the Moon, commenting, "I had to constantly turn my head around, because we felt we were not alone there. We had no choice but to pray." Later, after becoming a UFO proponent, Mitchell commented, "I have no doubt that extraterrestrials could very well have populated or made structures on the far side of the moon."

Mnaseas of Patrae

A prominent Greek historian of the late third century BCE. Mnaseas of Patrae spoke of a time when the Earth had no moon. He stated that the aborigines known in the land as the **Arcadians** existed before the Moon and were thus referred to as Proselenes, which means "people before the moon." His statement has been used as evidence of there having been a time when the Earth was without a moon.

Modern Day Mysteries

During the past twenty-five years, several ufologists and researchers have delved into examining old photographs of the Moon. They have allegedly discovered what looks to be artifacts on the Moon. While independent lunar researchers insist that the discoveries are natural formations and /or a trick of the lighting, researchers have pushed forward to prove their claims with photographs from NASA, websites and books. In addition, people have come forward with stories that they claim are real, that tell of a massive cover-up when it comes to this information. Among some of the objects allegedly found in the photographs are **ruins** of cities, a **bridge**, roads, pipelines, **domes**, spacecraft, rows of buildings, **monuments,** a plane-shaped **UFO**, a **cylindrical-shaped** UFO, a landing pad, a satellite dish, a skull and more. The conclusion from some researchers is that there was either once a civilization or a base located on the Moon sometime in the ancient past.

Moltke (Crater)

An impact **crater** positioned close to the southern rim of the **Sea of Tranquility** (Mare Tranquillitatis). A comment by astronaut Michael Collins on the Moltke **crater** has been interpreted as meaning there are roads on the Moon, near the crater. According to an *Apollo 11* transcript, Collins once commented, "Oh God, look at the Moltke; he's my favorite… You see all those roads – triangular roads leading right past him?"

Mona Lisa
See **Apollo 20 Hoax**.

Monks Mystery

On June 18, 1178, five Monks witnessed an eerie event on the Moon. What it was that they saw has been debated for years. The monks stated that they observed a sliver of a **crescent moon**, sitting low in the sky. Between the **horns** of the crescent on the lighted side, they noticed what appeared to be an explosion, complete with sparks and fire. Subsequently, smoke covered the Moon. It became dark and the Moon appeared to wither and writhe like a wounded snake. It later returned to its normal appearance. The story was recorded by the prominent English chronicler Gervase of Canterbury. Some believe that what the monks witnessed that day was a miraculous event, a sign to the monks from the divine. In the 1970s, as pictures of the Moon's **dark side** came to light, an astronomer named Jack Hartung suggested that a young crater named Giordano Bruno, may have been created by a large meteorite hitting the Moon during the twelfth century. It was thought that perhaps this is what the monks had witnessed. However, it was discovered later that the Giordano Bruno crater was actually over a million years old. Therefore, the twelfth century meteorite theory was discounted. Today, it is theorized that what the monks observed was a meteor exploding in Earth's upper atmosphere, positioned above the moon. The Moon writhing like a snake is said to have come from atmospheric instability from ascending summertime heat.

Mont Blanc

A prominent lunar mountain, located in the Moon's Montes Alpes range. It was named by famed astronomer **Johann Schroeter**. On September 26, 1789, **Schroeter** observed an anomalous speck of light at the foot of Mont Blanc for fifteen minutes. He reported that it had a brightness of a 5th magnitude star.

Monument(s)

Two different sets of mysterious structures made of stones and arranged in geometrical patterns, were found in different areas of the Moon in 1966. The first set was found by the Soviet spacecraft ***Luna 9*** in the Ocean of Storms. In an article from 1970 in *Argosy* magazine, Dr. A. Bruenko, a Russian engineer, reported: "There does not seem to be any height or elevation nearby from which the stones could have been rolled and scattered into this geometric form." The second set of structures was located by the United States unmanned craft ***Lunar Orbiter 2***. These structures too were arranged in perfect geometric patterns. *Lunar Orbiter 2*, a NASA spacecraft sent to the Moon on November 6, 1966, located the second set of what appeared to be shadows of monuments over the **Sea of Tranquility**. *Lunar Orbiter 2* took a picture of six shadows of structures shaped like monuments and also arranged in geometric shapes. Because *Lunar Orbiter 2's* cameras were facing downward towards the **obelisks**, only their silhouettes could be seen. Scientists estimated the largest structure to be fifty feet wide at the bottom and from forty to seventy-five feet in height. Russian scientists that studied the *Lunar Orbiter 2* photographs placed them at three times that height, which would be the equivalent of a building standing nearly fifteen stories high. The silhouettes captured by *Lunar Orbiter 2* have been described as having the appearance of the Washington Monument. Others likened them to the great **pyramids** of Egypt. These figures have precise shapes and patterns and are perfectly aligned in geometric positions. They have been dubbed the "**Blair Cuspids**," after anthropologist William Blair who examined the photographs taken by *Lunar Orbiter 2*. His comments about the objects appeared in the *Washington Post* on November 22, 1966. The title of the article was "Six Mysterious Statuesque Shadows Photographed on the Moon by Orbiter." The article was additionally published in the *LA Times*. Among his comments about the shadows, Blair stated, "If the cuspids really were the result of some geophysical event it would be natural to expect to see them distributed at random. As a result, the triangulation would be scalene or irregular, whereas those concerning the lunar object lead to a basilary system, with coordinate x,y,z to the right angle, six isosceles triangles and two axes consisting of three points each." Soviet Space engineer Alexander Abramov examined pictures of the monuments, determined their positions and concluded that they formed an Egyptian triangle much like the arrangement of the great pyramids of Egypt.

Moon Base

There have long been rumors that there is a secret base on the Moon placed there and operated by a government or governments of Earth. This hypothetical base is believed to be filled with personnel from Earth that were trained and sent to live on the Moon and establish a **colony**. Some believe that this base is operated only by humans. Others contend that humans and **extraterrestrials** are working together on the Moon. In 1994, NASA's *Clementine* spacecraft took pictures of what some believe is an artificial construction. The figure is shaped like a V and is located on the **far side** of the Moon in Mare Moscoviense (a mare that is positioned in the Moscoviense basin). It appears to be made up of two rows of seven lights. The design is symmetrical and measures roughly 500 ft by 420 ft. To some this resembles a base and they consider it proof that a Moon base does exist. In his book *Nothing in This Book Is True, But It's Exactly How Things Are* (page 179), author Bob Frissell talks about a secret government that has established a base on the Moon. He writes, "First they made a base on the Moon, using it as a satellite to go deeper into space. They built three small bubble type cities on the dark side. There was an accident on one of these and many people were killed. Records will indicate that there have been more than 2,000 secret missions to the Moon."

Moon Conspiracy Theories
See **Conspiracy Theories**.

Moon Dwellers

Alleged inhabitants of the Moon. See **Selenites**.

Moon Goddesses
See **Aega; Aine; Anahita; Anunit; Arianrhod; Artemis; Candi; Chang-E; Gleti; Hecate; Heng-O; Hina; Jyotsna; Lucina; Luna (2); Mama Quilla; Marama; Sadarnuna; Sarpandit; Selene; Tanit; Teczistecatl; Zirna.**

Moon Gods

See **Khonsu; Klehanoai; Mah; Mani; Men; Nanna; Thoth**.

Moon Hoax
See **Great Moon Hoax**.

Moon Illusion

An optical illusion that has been baffling scientists since around the fourth century BC. It is when the Moon appears larger when low near the horizon and smaller when elevated in the sky. The reason for this misperception is still being debated today.

Moon Landing(s)
See ***Apollo 11; Apollo 12; Apollo 14; Apollo 15; Apollo 16; Apollo 17***.

Moon Landing Conspiracy Theory

The hypothesis that the Apollo Moon landings were faked. Also known as the Moon landing hoax, it is one of the most well-known conspiracy theories in the United States. According to supporters of the theory, the Moon landings never took place. In addition, some believe that the images of the landings, as well as the **astronauts** walking on the Moon, were shot in movie studios or elsewhere on Earth. Some of the evidence that supporters use to back their theory is a lack of stars in the pictures, hair on the camera lens and a waving flag on the Moon. In his book *The Secret Influence of the Moon* (page 27), author Louis Proud explains why there are no stars seen in the Moon photographs stating, "There is a simple reason why they aren't: glare created by the reflective lunar **surface** (rocks especially). If you wanted to perceive the stars while standing on the **surface** of the Moon during the lunar day, you would have to shield your eyes for a moment to allow them to adjust to the darkness. During the lunar night there would be no light to create glare, making the stars stand out like floodlights. All of the **Apollo** missions took place during the lunar day—hence, the absence of stars in photographs."

In another version of this conspiracy theory, there are some who believe that the astronauts did visit the Moon; however, some of what they witnessed while there was covered up. It is thought that some of the film footage released was real and the other part was recorded in a studio or a selected film location. It is believed that the second portion of the film was created to keep the truth of what the astronauts had seen on the Moon from the public. They believe that artifacts, ruins and structures on the Moon were not acceptable for public viewing; therefore, footage was reshot of the Moon landings on Earth and then presented to the public.

In one fantastical story, it is said that when ***Apollo 11*** landed on the Moon, numerous **UFOs** were present. It is thought that the beings inside the UFOs wanted to see the visitors and possibly monitor their behavior. It was determined that those images especially could not be shared with the public. Another take on the Moon landing conspiracy theory states that the *Apollo 11* mission was faked to trick the world into believing that the United States was the first nation to land on the Moon. NASA reportedly has cleared up most of the questions surrounding the missions and the Moon landing conspiracy theory has since lost most of its following.

Moon Landing Hoax

See **Moon Landing Conspiracy Theory**.

Moon Light

See **Brighter Moon**.

Moon Missions to Locate Ruins Conspiracy Theory

The theory that the real reason for the missions to the Moon was to secretly look for remnants of an ancient civilization.

Moon Pigeons

Small white **UFOs** in the Moon's vicinity that were witnessed by NASA **astronauts**.

Moon Pillar

A type of lunar halo. It has the shape of a long beam of light reaching from above the Moon to below it. It is rarely seen, as it can only be observed when the Moon is close to the horizon.

Moon Placed in the Sky for Earth Theory

The theory that **extraterrestrials** brought the Moon to Earth's orbit for the benefit of life on the planet and to act as a timepiece for mankind.

Moon Rocks

Apollo **astronauts** brought back a total of 2,196 rock samplings taken from the lunar

surface. Their weight totaled 382 kg (842 lb). Scientists have found some samples to be 4.5 billion years old. This is enigmatic because that age makes them one billion years older than the oldest rocks found on the Earth. It also means that they are nearly as old as the solar system. After testing some of the samples, scientists found them to have processed metals in their composition. The elements found include brass, mica, neptunium 237 and uranium 236. These metals are not known to occur naturally. Additionally, amphibole and near-pure titanium were detected in some. Researchers are perplexed as to how they came to be on the Moon. Additionally, the rock samples taken from the Moon were also found to be magnetized. This is yet another mystery since the Moon has no magnetic field. Where the magnetism came from cannot be explained. One theory to explain these anomalies is that some of the Moon rocks may have originated somewhere other than the Moon. See **Color; Genesis Rock; Orange Soil.**

Moonbeam

A ray of light emanating from the Moon.

Moondust

A dusty dry element of lunar soil. Moondust has been described as being powdery, a white to yellow color and smelling like gunpowder. An article from NASA titled "Apollo Chronicles: The Mysterious Smell of Moondust" (January 30, 2006), tells what moondust is made from stating, "Almost half is silicon dioxide glass created by meteoroids hitting the moon. These impacts, which have been going on for billions of years, fuse topsoil into glass and shatter the same into tiny pieces. Moondust is also rich in iron, calcium and magnesium bound up in minerals such as olivine and pyroxene." Initially, before any of the **Apollo** missions went to the Moon, it was believed that since the dust had accumulated for billions of years, the spacecrafts might just sink when landing on the Moon. Mysteriously, the moondust turned out to not be deep at all. This has baffled scientists. The Moondust was clingy, and annoyed the astronauts as they sometimes found themselves covered in it. It was often carried into the Lunar Modules with them on their spacesuits after working on the lunar **surface**, as it was difficult brush off. But once in, the astronauts were able to examine it closer. That's when they were able to touch it, sniff it and in one instance, taste it. Even though one of the astronauts turned out to be allergic, the moondust seemed harmless to their systems. There were various comments from Apollo astronauts about Moondust. In "Apollo Chronicles: The Mysterious Smell of Moondust," two Apollo astronauts are quoted on their experiences with Moondust during their missions. "It's soft like snow, yet strangely abrasive," said **Eugene Cernan** of ***Apollo 17***. Cernan also stated, "It smells like spent gunpowder." After tasting it onboard the spacecraft, John Young of ***Apollo 16*** commented, "not half bad."

Moonless Earth

See **Pre-Lunar Earth**.

Moonlight

In Earth's nighttime sky, it takes light that is shining from the Moon approximately a second and a half to reach Earth.

Moonport

An installation from which spaceflights are sent to the Moon.

Moonquakes

Rumblings on the Moon that are similar to earthquakes. Moonquakes were originally detected by the Apollo **astronauts**. They can last from a few minutes to up to an hour and sometimes more. Four types of moonquakes have been identified. The first is the cavernous moonquake. These occur deep beneath the lunar

surface at around 435 miles (700 km). There are meteoric moonquakes that come from meteorites crashing into the Moon. Thermal moonquakes also occur, which are caused by the Sun's heat. However, it is the shallow moonquakes, which register around 5.5 on the Richter scale, that are considered to be the most dangerous. Scientists are not one hundred percent sure of all of the reasons that the Moon occasionally rumbles.

Moonscape

The lunar **surface** or a depiction of it.

Moonshot

The operation of sending a spacecraft to the Moon.

Moonstone

A shimmering, pearly white, semiprecious gem that exudes a mysterious, unearthly glow. Moonstone is from the feldspar mineral category. It is said to contain healing properties. In days of old, it was believed to be the manifestation of the Moon's rays.

Moonwalk

A walk taken by an **astronaut** on the lunar **surface** for exploratory and or scientific purposes. Twelve people have walked on the Moon. They include **Neil Armstrong, Buzz Aldrin, Alan Bean, Charles Eugene Cernan, Pete Conrad, Charles Duke, James Irwin, Edgar Mitchell, Harrison Schmitt, David Scott, Alan Shepard and John Young**.

Moore, Patrick
(Sir Patrick Alfred Caldwell-Moore)
1923–2012

A British astronomer, lunar researcher and writer. He was also a radio and television host. Moore famously authored more than seventy astronomy books, was a presenter for the BBC's *The Sky at Night* and was President of the British Astronomical Association. He was President (and co-founder) of the Society of Popular Astronomy. Moore once witnessed a peculiar light in the **Gassendi crater** that he could not explain. He stated that the color of it was "reddish" and it appeared to be glowing. He described what he saw as "a color phenomena." He observed it from April 30 to May 1, 1966. Moore was also witness to the famous "**bridge**" on the Moon; a twelve mile structure that appeared and then after a long while, mysteriously vanished.

Mountains
See **Lunar Craters**.

Music
See **Space Music**.

N

Names of the Moon

Chinese (Yuet); Dutch (Maan); Danish (Mane); Egyptian (Pooh); Eskimo (Tatkret); Finnish (Luna); French (la Lune); German (der Mond); Greek (Menos); Hawaiian (Mahina); Irish (Gealach, Luan); Italian (la Luna); Japanese (Tsuki, common); Japanese (Otsukisama, reverent); Korean (Tal); Persian (Mah); Peruvian (Mama Quilla); Polish (Ksiezyc); Polynesian (Hina); Portuguese (a Lua); Romanian (Luna); Russian (Luna); Serbo-Croatian (Mjesec); Spanish (la Luna); Turkish (Ay); Welsh (Ileuad).

Nanna

A Moon god in Mesopotamian (Sumerian) beliefs. Nanna illuminated the night as he traversed the heavens in his chariot.

NASA Astronaut(s)
See **Astronauts**.

NASA Technical Report TR-277
See ***Chronological Catalog of Reported Lunar Events.***

National Aeronautics and Space Administration (NASA)

An agency that directs the civilian space program. It is also responsible for the aeronautics and aerospace research areas. In addition, it is an autonomous agency of the United States Federal Government. See ***Chronological Catalog of Reported Lunar Events;*** **Astronauts;** ***NASA Technical Report TR-277*****;** ***Apollo; Apollo 7; Apollo 8; Apollo 9; Apollo 10; Apollo 11; Apollo 12; Apollo 13; Apollo 14; Apollo 15; Apollo 16; Apollo 17; Apollo Program.***

Nazca Lines

Mysterious lines were found on the Moon that are similar to the Nazca lines on Earth, leaving some to speculate that they are connected.

Nearside

The side of the Moon that is facing the Earth. The back side is known as the **far side**.

Nelson, Buck
1895–1982

An American Ozark farmer and extraterrestrial **contactee**. Nelson's **extraterrestrial** contact was referred to as "Little Bucky." Nelson claimed that he was an expatriate residing on Venus. In 1955, Nelson claimed to have journeyed to the Moon with his extraterrestrial companion. Once there, he rested, observed a building and witnessed children playing with various sized dogs, before leaving on another cosmic trip.

Nibiru
See **Anunnaki**.

Nicholas of Cusa
1401–1464

Famed German philosopher and astronomer from the Middle Ages. Nicholas of Cusa was one of the many astronomers in history who believed the Moon is inhabited.

Noah's Ark

There is a theory that the Moon may be a rescue ship similar to the biblical Noah's ark. It has been speculated that the Moon was brought into our solar system by beings that had escaped a disaster on their home planet and are still living inside the Moon. Since the **hollow moon theory** first came to light, there have been many speculations about the Moon's origin. A civilization living inside the Moon is but one idea. It is thought that beings in a spaceship, if well enough equipped, could have hurled themselves into space, on a cosmic journey that could last for many years. The idea of a hollow moon became popular after the ***Apollo 11*** Lunar Module was crashed into the Moon for testing. The Moon reverberated for a long period of time. There is also the evidence that the Moon's crust is made up of some unusual substances and metals. Researchers have indicated that this gives the Moon a protective outer layer, which encourages the speculation that there is something more going on with Earth's **satellite** than meets the eye. In contemplating the idea that the Moon could be a "Noah's ark," researchers have considered the possibility that the Moon could one day become a true ark for Earth inhabitants if there were a nuclear war or celestial event that threatened to wipe out life on Earth. Dr. Bernard H. Foing of the European Space agency commented on this in an article from *The Daily Galaxy,* "Earth's Secrets to be kept in a Lunar 'Noah's Ark.'" Dr. Foing stated, "If there were a catastrophic collision on Earth or a nuclear war, you could place some samples of Earth, including humans on the Moon. You could repopulate the Earth afterwards, like a

Noah's Ark."

Number Matrix

Per Christopher Knight and Alan Butler, the authors of ***Who Built the Moon,*** there is a mysterious number pattern that connects the Moon, Earth and the Sun.

O

Obelisks

The *Merriam-Webster Dictionary* defines an obelisk as a "4-sided usually monolithic pillar that gradually tapers as it rises and terminates in a pyramid." Many obelisks are located on Earth and allegedly some have been discovered on the Moon. The original reason behind the building of obelisks remains unknown. The fact that some may have been located on the Moon is even more mysterious. Many question the connection between those found on the Earth and the constructions on the Moon. In an article titled "Mysterious 'Monuments' on the Moon," *Argosy Magazine,* August 1970, Volume 371, Number 2, biologist, writer and paranormal researcher Ivan T. Sanderson asked, "Is the origin of the obelisks on this Earth and those on the moon, the same? Could both be ancient markers originally erected by alien space travelers for guidance of later arrivals?" See **Blair Cuspids**; **Monuments**.

Objects Crossing the Moon

Astronomers from various areas around the world have witnessed strange objects moving across the Moon as far back as the 1800s. There is no logical explanation as to what these baffling objects are. In 1820, in the midst of a **lunar eclipse**, French astronomers reported seeing mysterious objects moving in straight, uniform rows and traveling away from the lunar **surface**. Similarly, in 1869, three moon watchers from the United States witnessed numerous objects traveling across the Moon in straight, parallel lines in what appeared to be a deliberate configuration. In 1874, a lunar observer from France reported seeing several black objects traversing the lunar floor. Additionally, an astronomer from Czechoslovakia in that same year saw a luminous, white object moving at a slow pace across the Moon. It eventually flew off into space. In 1892, a Dutch astronomer spotted a dark round object moving horizontally across the Moon. In 1896, William R. Brooks, an astronomer at the Smith Observatory in America, also witnessed a dark round object moving across the Moon. He stated that the object was one-thirtieth of the Moon's diameter. He also stated that the object was moving at great speed, crossing the moon in approximately three or four seconds. In 1899, two astronomers in Arizona witnessed an illuminated object moving across the Moon, close to the lunar floor. In 1912, an English astronomer by the name of F.B. Harris reported seeing a massive dark object measured at 250 miles long and 150 miles wide on the Moon. Harris later stated that what he had seen could have been the shadow of something enormous crossing above the Moon. Of this account, Harold T. Wilkins wrote in his book *Flying Saucers on the Attack* (page 219), "It was like a crow poised and I think a very interesting and curious phenomenon occurred on that night!" In another odd account, in 1912 during a lunar eclipse, astronomers located in France and Britain, saw what they described as a "superb rocket" burst forward from the lunar surface. To date, there have been no explanations on what the objects crossing the Moon are, nor their origin.

Oceanus Procellarum

A large lunar mare on the nearside of the Moon that has at least two mysteries surrounding it. Purportedly, a strange craft-like object was found in the region of Oceanus Procellarum. The Soviets' *Luna 13* (unmanned craft) set down in the vicinity of Oceanus Procellarum on December 24, 1966. Reportedly, in the photographs taken on that mission, a perplexing object is seen. The object is described as lying on the Moon's **surface**. It appears to have wheels that are connected to an axel. Researchers have speculated that this odd item is a piece of debris from *Luna 13* itself. However, what the pictures revealed reportedly does not resemble any piece of the *Luna 13* craft. In addition, the strange item looked nothing like pieces to the previous crafts that had landed on the Moon. In another strange tale, in the area of Oceanus Procellarum there is an alleged lunar city. Images and analysis of the city are given by an online researcher known as Copernicus2013. His work on Oceanus Procellarum and enhanced images of the city can be viewed at *A City on the Moon, Oceanus Procellarum* https://www.youtube.com/watch?v=d2uTfo03amI.

O'Neill, John Joseph
1889–1953

Journalist and former science editor of the *New York Herald Tribune*, an aerial phenomenon expert and Pulitzer Prize winning astronomer. In 1953 O'Neill reported that he had seen a sizable **bridge** on the Moon stretching above **Mare Crisium,** and that it seemed to be artificially made. Other astronomers reported having seen the structure as well. It was estimated to have been twelve miles long. The bridge is said to have appeared out of nowhere and eventually, just as mysteriously, disappeared. Since O'Neill first reported it, the bridge was named O'Neill's Bridge.

O'Neill's Bridge
See **Bridge; John O'Neill**.

Operation Moon-Blink

An investigation from the 1960s, where NASA requested observatories from around the world to monitor and photograph the Moon. Within just a few months of Operation Moon-Blink being created, twenty-eight TLPs (**Transient Lunar Phenomenon**) were recorded.

Orange Ray

There was a mysterious tale that appeared in *The Marine Observer,* 11:49, in April, 1934 and retold in the *Reader's Digest* book *Mysteries of the Unexplained* (page 254). The curious account tells of passengers aboard a ship named the *S.S. Transylvania* that was located in the North Atlantic at the time. Moon observers that evening had seen auroras several hours before the incident. Later they witnessed what they describe as an "orange ray" shooting upward from the Moon. The odd effect lasted for 15 minutes. This occurred on the night of May 2, 1933.

Orange Soil

The Astronauts of ***Apollo 17*** discovered orange soil on the Moon. Astronaut Jack Schmitt, who was the geologist on *Apollo 17*, recognized that an orange color in rocks meant that rust was present. This was perplexing to him because without the presence of iron, water, oxygen and free hydrogen, there can be no rust. The Moon is free of those materials.

Orbit

The Moon orbits the Earth counterclockwise when viewed from the north celestial pole. It also rotates around its axis counterclockwise. The Earth has the same orbital direction around the Sun and rotational direction around its axis.

At an orbital speed of 2286 miles/hr (about 3679 km/hr), the Moon's synodic orbital period around the Earth is about 29.5 Earth days, whereas its sidereal orbital period around the Earth is about 27.3 Earth days (see **Lunar Days**).

Orbiter 2
See ***Lunar Orbiter 2***.

Orbiter 3
See ***Lunar Orbiter 3***.

Orbs
See **Luminosities**.

Origin

The origin of the Moon remains a mystery. Planetary scientist William Hartmann states in his book *Origin of the Moon*, "neither the Apollo **astronauts**, the Luna vehicles, nor all the king's horses and all the king's men could assemble enough data to explain the circumstances of the moon's birth." There are many theories as to how the Moon came to be in Earth's orbit. Several are creation-based and some are more esoteric. There are five "popular" theories that have been put forward by scientists as to the origin of the moon. All have been discounted. The most accepted of all the theories is the capture theory. The five theories are: 1) the **capture theory**, 2) **giant impact theory,** 3) **double big whack theory**, 4) the **coaccretion theory** and 5) **the fission theory**. See **Planetoid Theory; Spaceship Moon Theory.**

Orion's Belt
(Belt of Orion)

A prominent group of bright stars within the Orion constellation. The stars in Orion's Belt include Alnitak, Alnilam and Mintaka. They are also referred to as the "Three Sisters" or the "Three Kings." In a mysterious tale, it has been said that certain sites visited on the Moon by the **astronauts** were chosen based on the position of Orion's Belt, the **Sea of Tranquility** (Mare Tranquillitatis), the landing site for ***Apollo 11***, being one of them. Some believe that the Sea of Tranquility was chosen based on specific dates and times, as well as the altitude and longitude, so that on the date of the landing, the coordinates would line up directly under Orion's belt.

Orthostat 47
See **Lunar Map at Knowth**.

Ovid
(Publius Ovidius Naso)
43 BC–17 AD

A noted poet best known for his work *Ars amatoria* and *Metamorphoses*. Ovid mentions in his writings a time when the Earth had no moon. Ovid wrote in his work titled *Fasti*, "The Arcadians are said to have possessed their land before the birth of Jove and the folk is older than the Moon." His words are seen as evidence, by conspiracy theorists and **UFO** proponents alike, that the Moon is not a natural **satellite**, but was created elsewhere and brought into our Solar System.

P

Paperclip

An odd anomalous construction located on the lunar floor as discovered in a NASA photograph by writer and researcher **Richard Hoagland**. Hoagland named the object the "Paperclip." The paperclip appears to be sitting on a pole and appears to be a part of some sort of assembly. Some lunar researchers believe it to be an unnatural object, while others are skeptical. It is purportedly immense.

Paracelsus C

An impact crater located on the dark side of the Moon. In images taken by ***Apollo 15*** astronauts and a Lunar Reconnaissance Orbiter, there are two constructions in Paracelsus C that have been found by lunar researchers. They appear to be walls positioned on the crater floor.

After careful study of the pictures, researchers maintain that they are walls that seem to form a passageway. It is believed that this passageway leads beneath the Moon's **surface**. Located nearby are sites that appear to have been excavated by unknown means.

Pareidolia
See **Lunar Pareidolia**.

Pelasgians

They are known as the first inhabitants of Greece and the people that lived during pre-lunar Earth. See **Arcadians; Pre-lunar People; Proselenes**.

Piccolomini (Crater)

A prominent impact **crater**. On the night of July 6, 1954, astronomer **Frank Halstead** (a former curator at the Darling Observatory in Minnesota) witnessed a straight black line in the Piccolomini crater. It was an area that had been observed in the past, yet this was the first occasion that the anomaly had been spotted. It was witnessed not only by Halstead but by seventeen other people. The strange line soon disappeared.

Pickering, William Henry
1858–1938

American astronomer who made several significant Moon observations. Pickering is also well-known for establishing a number of observatories including the prominent Lowell Observatory in Flagstaff, Arizona. It was Pickering that discovered Phoebe (on March 18, 1899), Saturn's ninth moon. He was an avid Moon observer. Pickering investigated lunar **craters** and carefully watched the changes in the **Eratosthenes crater**. Pickering reported seeing "traveling dark objects" that were traversing the lunar floor. He wrote of this odd occurrence: "In trying to find conclusive arguments for or against the existence of animal life upon the Moon, I have necessarily studied not only the routes along which it appears to travel, but also the reasons for which it might be expected to travel." Pickering concluded that what he had witnessed were a form lunar insect. He reported that they covered 20 miles in 12 days. Pickering believed that these lunar insects were the cause of the changes that he found in the Eratosthenes crater. In his book *Our Mysterious Spaceship Moon,* Don Wilson writes, "no one questions Pickering's integrity and competence. He did see something. What it was remains a mystery."

Pitatus (Crater)

A primordial impact crater that is situated near the lower part of Mare Nubium. In the 1600s, two mysterious anomalies were reported in Pitatus by famed astronomer **Gian Domenico Cassini**. The first occurred on November 12, 1671. On that day, Cassini witnessed what appeared to be an unexplained small, light-colored, cloud above the crater. On October 18, 1673 Cassini observed an unexplained white glowing spot of light.

Planetoid Theory

The theory that the Moon is a planetoid that was hollowed out thousands of years ago, in a place outside of our galaxy. See **Hollow Moon Theory; Soviet Union; Spaceship Moon Theory**.

Plaque

A plaque was left on the Moon by the ***Apollo 11*** astronauts **Neil Armstrong** and **Buzz Aldrin**. On it is inscribed: "Here men from the planet Earth first set foot upon the Moon, July 1969, A.D. We came in peace for all mankind." The plaque also holds the names of Neil Armstrong, Buzz Aldrin, **Michael Collins** and President Nixon.

Platform
See **Archimedes Platform**.

Plato
See **Atlantis; Plato Crater**.

Plato (Crater)
A prominent impact crater named after the famous Greek philosopher **Plato**. It is also the darkest area of the Moon. Even so, the Plato crater is regarded as one of the "**UFO** hotspots" on the Moon. Numerous brilliant lights have been observed there over the years. The extraordinary lights come in geometrical patterns, grids, triangles and more. On April 15, 1948, noted astronomer F.H. Thornton momentarily saw a luminous flash of light on the floor of the Plato crater. He described its color as orange-yellow and stated that it was similar to the flash of an anti-aircraft shell exploding from a distance of nearly ten miles away. In his book *Elder Gods of Antiquity* (page 102), author M. Don Schorn writes of the odd light witnessed by Thornton stating, "The light was definitely not from a falling meteor or a volcanic explosion and no explanation for that phenomenon was ever given."

During the late 1860s, various patterns of light materialized in Plato and were seen on and off for the next decade. They appeared so often that selenologists allocated numbers to them. They also kept track of the intensity of the lights. The reputable selenographer W.R. Brit amassed over 1,600 anomalous lights accounts. The collection can be found in the Royal Astronomical Association's library. Mysterious lights, **transient lunar phenomenon** and strange activity in Plato include:

1685, December 10, a red streak was seen on the crater floor during a **lunar eclipse**.

1751, April 22, a yellow streak of light was seen across the darkened floor.

1788, January 11, a brilliant light was seen on the dark side.

1824, December 8, a bright spot was seen in the southeast.

1870, May 13, a spectacular display of lights was witnessed.

1872, July 16, the northeast area of the floor appeared hazy.

1878, October 5, a thin, shimmering, light colored cloud was seen.

1882, March 27, the floor glowed with a white light.

1887, February 1, a bright light was seen.

1895, May 2, a streak of light was observed for twelve to fourteen minutes.

1907, January 22, a luminosity was witnessed in part of the crater.

1937, December 12, an intense streak of orange-brown, was seen on the eastern wall.

1938, January 16 to 17, the floor was covered with a gold veined pattern.

1938, February 14, a golden-brown spot was witnessed on the east wall. It had a yellowish glow to it and traveled down and over the floor.

1945, October 19, a brilliant flash of light was seen on the floor close to the eastern wall.

1952, April 4, an obscuration of the crater floor was witnessed.

1964, May 20, a strong reddish-orange color was observed for ten minutes on the western rim.

1964, July 17, dim pink tinted areas were seen at the bottom of the west wall and also on the edge of the north wall.

1964, November 14, the peak on the west wall glowed brightly; the foot of the peak had a blue glare. In the southwest, there was a small, red glow.

1966, August 4 to 5, a red area was seen on the northeast wall and floor. The first day it was seen for fifty-three minutes; the second, twenty-six minutes.

1966, December 23, numerous blinking lights were witnessed on the floor the crater. This activity took place for 55 minutes.

See **Luminosities.**

Pleiades
See **George Adamski; Alaje of the Pleiades; Billy Meier**.

Plutarch
ca. 45–120 CE

A prominent Greek philosopher, biographer and essayist. He was best known for his works titled *Parallel Lives* and *Moralia (Moral Essays)*. He also wrote *On the Face in the Orb of the Moon*. Plutarch believed the Moon to be a smaller version of Earth, complete with peaks and valleys akin to a mountainous terrain. He also believed there were strange demon-like creatures inhabiting the Moon. In his work *The Roman Questions,* he wrote of a civilization that lived prior to there being a moon in the sky stating, "There were Arcadians of Evander's following, the so-called pre-Lunar people."

Pluto (Crater)
See **Harold T. Wilkins**.

Pre-Lunar Earth

According to ancient writings, oral legends and tales from various cultures, there was a time in Earth's history when there was only the Sun and stars that appeared in the sky; a time before there was a Moon. Modern-day science teaches that the Earth cannot survive without the Moon, that all life would die without it. Still, the stories persist. It is one of the Moon's most intriguing mysteries. There are credible people from ancient times that spoke of a period when there was no moon in the sky. These were men of prominence who helped shape mankind's view of the world and the universe. People who wrote about a time when there was no moon include: **Anaxagoras of Clazomenae**, **Aristotle**, **Giordano Bruno, Censorinus**, **Democritus**, Dionysius Chalcidensis, **Hippolytus of Rome**, **Lucian of Samosata**, **Mnaseas of Patrae**, **Plutarch**, **Ovid**, Stephanus of Byzantium and **Theodorus of Cyrene**. Some gave the name of a culture that lived during that period. They were known as the **Arcadians** and were sometimes referred to at the Proselenes. Biblical writers also spoke of a time before the Moon. The **Bible** book of Job 25:5 states, "the grandeur of the Lord who 'Makes peace in the heights' is praised and the time is mentioned before [there was] a moon and it did not shine." In Psalms 72:5 it states, "Thou wast feared since [the time of] the sun and before [the time of] the moon, a generation of generations." Cultures that have tales of a Moonless sky include the **Bogota** people of Columbia, the people of Bolivia, the **Mayans** and the **Tibetans**. See **Gondwana.**

Pre-Lunar People
See **Arcadians; Pelasgians; Proselenes**.

Proclus (Crater)

A small, young impact **crater** that dates back to the Copernican age. A British astronomer by the name of C. Barrett once witnessed a brilliant light in the Proclus crater. When conveying what he had seen, Barrett reportedly was adamant that what he had observed was not from the Sun's reflection as was suggested by others. On September 8, 1954, an anomalous blue light was seen within the crater by astronomer Valdemar Axel Firsoff. The brightness of the light varied in intensity.

Project A119
(A Study of Lunar Research Flights)

An alleged clandestine strategy developed in 1958, in which the United States Air Force would explode a nuclear bomb on the lunar **surface** in an effort to answer questions relating to planetary astronomy and exogeology. The project was never executed, purportedly due to fear of a negative response from the public and foreign countries.

Project Gemini

Project Gemini was a prelude to the **Apollo**

missions to the Moon. It was designed to be a link between the Mercury and **Apollo** programs. The primary mission of Project Gemini was to train astronauts, test equipment and perform assessment of all mission procedures that were to be used in future **Apollo** missions. A total of ten crews went up on assigned missions under Project Gemini. Each mission carried two astronauts as it orbited Earth. The missions were performed in 1965 and 1966. Missions lasted anywhere from five hours to 14 days. The astronauts in each of the twelve Gemini missions sighted one or more **UFOs**. These UFOs have never been identified and remain a mystery. UFO researchers believe that the missions were being watched by advanced beings that had an interest in the space program's Moon agenda. See ***Gemini 1; Gemini 4; Gemini 7; Gemini 9A; Gemini 10; Gemini 16***.

Project Horizon

A proposed project from 1959 for the United States Government to place a station on the lunar **surface**. Its main goal would have been to protect the United States' interests on the Moon.

Project Mercury

See ***Mercury 9***.

Project Moon-Blink

A plan created by NASA in 1965 to have unusual phenomena occurring on the Moon investigated. The work was done by Trident Engineering out of Annapolis, Maryland contracted by the Goddard Space Flight Center located in Greenbelt, Maryland.

Project Whiteout

(Whiteout Project)

An alleged planned mission to the Moon. The goal was to deliver equipment to the **surface** of the Moon between 1959 and 1964 for the purpose of setting up a lunar base. The base was to be named Luna. It was to have been a part of the *Clementine* project and Alternative 3. Mysteriously, it is said that both humans and extraterrestrials were to live there.

Prophet

Adam (the first man in Judaism, Christianity and Islam) was called a prophet of the Moon by some because he taught others about the Moon. He was considered to be an envoy of the Moon.

Proselenes

The name of an ancient civilization which, according to **Aristotle**, existed in Greece before there was a Moon in the sky. They are also known as the **Arcadians** and **Pelasgians**.

Psalms

See **Bible.**

Pyramid(s)

Constructions shaped as pyramids were photographed on the Moon by NASA's *Lunar Orbiter* probes during the 1960s. It is thought that pyramids on the Moon and those on the Earth may have a connection. This stems from the idea that there was once an ancient city on the Moon that was thought to have been destroyed during a **war**. To some researchers, what are believed to be artifacts and **ruins** on the Moon are from this period in the Moon's and Earth's galactic history. One of the cultures is thought to be that of **Atlantis**. Others believe a group of beings known as the **Anunnaki** may have been involved. It is therefore understood by some esoteric thinkers that the ruins on the Moon, along with the pyramids, are directly connected to highly advanced, intelligent beings that came into our solar system long ago, established colonies on the Moon, Earth and Mars, and eventually left. In his book *Ancient Aliens on the Moon* (page 188), **Mike Bara** writes about an anomalous construction in the **Tycho crater** that is shaped as a pyramid. He describes it as "very bright and appears to have a 4-sided pyramidal structure." See ***Apollo 12***; **Blair Cuspids; Howard Hill; Monuments; Ruins; Soviet Union; Structures; Triesnecker Crater**.

Q/R

Rabbit in the Moon

A lunar pareidolia image seen on the nearside of the Moon that appears to be a rabbit. The Mare Tranquillitatis is seen as the rabbit's head and Mare Nectaris and Mare Fecunditatis are the rabbit's ears. The tale of a rabbit on the Moon originated in Sri Lanka. In the story, the Buddha had lost his way in the woods. It was a rabbit that helped the Buddha. The Buddha was penniless and was unable to pay the rabbit for his kindness, and also informed him that he was hungry. The rabbit suggested that Buddha kill him, cook him and then consume him. Once Buddha created a fire, the rabbit leaped in. According to the tale, the Buddha retrieved the rabbit from the flames and put the rabbit on the Moon. This tale is told in similar forms in other legends as well. In Chinese lore, it was a rabbit by the name of Yutu that assisted the Moon goddess Chang'e to escape from her husband Hou Yi (after drinking the elixir of eternal life).

Radio Signals

Radio waves that are used to send and receive messages. Radio signals have been picked up in the area of the Moon. They were detected and reported in the Moon's vicinity in the years 1927, 1928, 1934, 1935 and 1958. In 1935, two scientists by the names of Stormer and **Van der Pol** picked up radio signals both on the Moon and in the area around it. In 1958, astronomers in America, Britain and the Soviet Union located a **UFO** that was heading in the direction of the Moon. It was reportedly travelling at a speed of 25,000 mph. Moon observers heard radio signals emanating from it. They were unable to decipher the signals.

Radius

The mean radius of the Moon is 1,080 miles (1738.092 km).

Railway

An odd formation found on the lunar **surface**. It appears to be a construction measuring approximately seventy miles long. Some lunar researchers believe it to be artificially made. It has been dubbed "The Railway." Some refer to it as the "Straight Wall." As time progressed, astronomers determined that what they were looking at was a large fault line. The fault line is pushing up through the lunar floor, which contributes to its mysterious appearance.

Rainbow UFO

See **China**.

Rang Like a Bell

A famous phrase that was used when the ***Apollo 12*** astronauts crashed the Lunar Module

The Aztec glyph for the Rabbit in the Moon.

into the **surface** of the Moon for seismic testing. To the surprise of scientists on Earth, the Moon reverberated for an extended period. It was said that the Moon "rang like a bell." See **Hollow Moon Theory**.

Rays
See **Orange Ray; White Rays**.

Record Album

The ***Apollo 11*** astronauts left a gold-plated 33–rpm record album titled *Camelot* on the Moon.

Refugee Theory

The theory that the Moon in the Earth's ancient past held refugees whose world was destroyed. See **Michael Salla**.

Remote Viewing
See **Ingo Swann**.

Reptilians

In **UFO** circles, the reptilians are a well-known race of malevolent aliens that some believe created the Moon eons ago. They are described as six to eight feet tall, bipedal, with scaly, green and sometimes brown reptilian skin. They have gold or sometimes yellow eyes resembling those of a cat. The Moon is believed to be a spacecraft created by these beings. They are thought to exist on the Moon (some believe they are living underground) and have enough advanced technology to influence mankind from there. It is believed that there is a conspiracy by the reptilians to enslave humanity. As the story goes, the reptilians depend on Earth for food and water for their survival. It is thought that resources from Earth are being extracted by these beings on a regular basis and that they do not want to be discovered. It is also thought that they use the Moon as a central location for mind control over mankind and have been exersizing this control for thousands of years. Some say that this extensive mind control is soon to end, as mankind awakens and becomes aware of its place in the universe.

Riccioli (Crater)

A prominent impact crater named after **Giovanni Battista Riccioli**, an Italian Jesuit priest, astronomer and lunar cartographer. Riccioli is roughly 87 miles (140 km) in diameter and boasts a variety of landforms. It is located south of the equator near the lunar western limb. Two mysterious occurrences have been recorded in Riccioli. On March 13, 1788, noted astronomer **Johann Schroeter** witnessed an odd bright spot of light located there. On the nights of September 28 and 29 1937, an area the crater was reported by one astronomer as glowing in an intense purple to amethyst hue.

Riccioli, Giovanni Battista
1598-1671

An Italian Catholic priest, astronomer and lunar cartographer. Riccioli named a number of lunar features and catalogued many of the larger craters. Riccioli's impression of the Moon was that of a dry and lifeless world. Unlike many of his peers, Riccioli did not believe there were Moon inhabitants, stating, "No man dwells on the Moon." The **Riccioli crater** was named in his honor.

Rima Hadley

Rima Hadley is a sinuous rille found west of where ***Apollo 15*** set down. It has been suggested that Rima Hadley is a volcanic vent from which lava spewed and produced some of the interesting characteristics of the area. The region has enormous boulders, described as being the size of a large house on Earth. There is a mysterious construction surrounded by a D-shaped wall found in the higher part of Rima Hadley, near where *Apollo 15* landed. It is thought to be a remnant of an ancient civilization once located on the Moon.

Roads

According to sources, what appear to be roadways have been located on the Moon. The pictures were taken by satellites. These mysterious roads are said to run through **craters**, hills and valleys.

Rocks

See **Genesis Rock; Moon Rocks**.

Rosetta Stone of the Planets

The Moon was dubbed "the Rosetta Stone of the planets" by American astronomer and planetary physicist Dr. Robert Jastrow, because of the complexities of understanding it and the fact that it is so old. A former *New York Times* science writer, Earl Ubell, expressed the same line of thought in his article "The Moon Is More of a Mystery than Ever" (April 16, 1972) when he wrote, "The Moon is more complicated than anyone expected; it is not simply a kind of billiard ball frozen in space and time, as many scientists had believed. Few of the fundamental questions have been answered, but the **Apollo** rocks and recordings have spawned a score of mysteries, a few truly breath-stopping."

Royal Astronomical Society of Great Britain

In 1869, the Royal Astronomical Society of Great Britain led an investigation into **anomalous lights** that lasted three years. The lights were most often seen near the Moon's **Mare Crisium** region. The lights were arranged in a variety of patterns including triangular and linear, were moving, and some appeared brighter at times than others. It was believed that these formations were the result of intelligent design and that thinking beings were behind the entire process. Some even looked for ways to decipher a possible message from the light patterns. The lights were also seen near the Plato **crater**. The society recorded nearly 2,000 accounts of these anomalous lights. The lights eventually disappeared.

Ruin(s)

Lunar researchers studying photographs of the Moon have allegedly found formations of what appear to be ruins of an ancient city. The renowned American writer and researcher Charles Fort once stated that in Europe, astronomers had asserted in scientific periodicals that they had observed ruins on the Moon in the past century. Conspiracy theorists believe that there is a cover-up when it comes to this information. Some hold that this is proof that there was once a civilization on the Moon and is evidence that we are not alone in the universe. Ruins that are allegedly on the Moon include **monuments,** partially destroyed walls, complex geometric shaped constructions, **structures**, **pyramids** and more that do not appear to be there by natural geological means. NASA expert geologist *Farouk* **El Baz** is quoted in a 1974 article in *SAGA Magazine* (Volume 47, Number 6, March 1974). He states, "We may be looking at **artifacts** from **extraterrestrial** visitors without recognizing them." In his book *Extraterrestrial Archaeology* (page 60), researcher and author David Hatcher Childress features an illustration showing a "layered outcrop of rock" that is approximately "8 meters tall," located close to where ***Apollo 15*** set down. Childress notes, "its similarity to the polygonal building style used at the massive wall of Sacsayhuaman in the Andes of Peru. Walls like this on Earth can be artificial." He asks the thought-provoking question, "Can they on the Moon as well?"

Russia(ians)

See **Soviet Union**.

Rutledge, William

A fake name assigned to a character from the **Apollo 20 Hoax**. According to the story, William Rutledge was a former NASA astronaut who went on a secret mission to the far side of the Moon. Once there, he and two others located ruins of an ancient city, a large cigar-shaped spaceship and an extraterrestrial, humanoid female pilot in stasis. He took video of the area as evidence and later uploaded it to the Internet. The story was later revealed to be an elaborate hoax. See **Apollo 20 Hoax.**

S

Sabaeans
See **Adam**.

Sacchari, Vito
In 1979, an engineer by the name of Vito Sacchari visited NASA's Houston offices along with a friend. The two requested to be allowed to view photographs of the Moon. They were let in after a discussion over whether it was allowed. The men were eventually given three days to review the pictures. However, they were not allowed to film, record, or even write about what they were seeing. They were also not to be left alone with the photographs. Sacchari studied 2,000 photographs of the Moon. The pictures allegedly contained large **extraterrestrial** structures, machinery on the lunar floor, a **UFO** and more. He also looked over Apollo missions transcripts. What Sacchari witnessed that day, he revealed to the public for their edification.

Sadarnuna
Sumerian Moon goddess. Sadarnuna reigns over the new moon.

Salla, Michael
1958–
Australian author and lecturer on exopolitics. Salla has indicated that the Moon is artificial. He believes that when it arrived into Earth's orbit, it held refugees from a planet that had been decimated. He also insinuates that there is a military industrial **extraterrestrial** complex on the Moon. The extraterrestrials that operate there are believed by Salla to be putting forward an agenda of assimilation for those on Earth. See ***Apollo 17***; **Stargate**.

Sample(s)
See **Moon Rocks**.

Sanderson, Ivan Terrance
1911–1973
American biologist, writer and paranormal researcher. He also wrote fiction under the pen name of Terence Roberts. In an article titled "Mysterious 'Monuments' on the Moon," *Argosy Magazine*, August 1970, Volume 371, Number 2, Sanderson, commented, "Many phenomena observed on the lunar **surface** appear to have been devised by intelligent beings." See **Obelisks**.

Santa Claus
The term Santa Claus was allegedly a secret code for Apollo **astronauts** when seeing **UFOs** or **extraterrestrial** spacecraft. See ***Apollo 8.***

Sarpandit
The Sumerian goddess of moonrise. Her name means "gleaming silver," symbolizing the reflective characteristic of the Moon.

Satellite
A celestial body orbiting a planet. The Moon is a satellite of the Earth. Exactly how the Moon came to be Earth's satellite is a mystery. The most popular and generally recognized hypothesis is the **Giant Impact Theory**. Some believe that the Moon is not a natural satellite.

Satellite Dish
See **Asada Crater**.

Schmitt, Harrison
1935–

Former NASA **astronaut**, geologist and former US Senator for New Mexico. Schmitt served aboard *Apollo 17*, the **Apollo program's** last mission to the Moon. His assignment was Lunar Module pilot. He is also the author of *Return to the Moon: Exploration, Enterprise and Energy in the Human Settlement of Space*. There are tales and rumors surrounding Schmitt and his comrades from *Apollo 17* that tell of strange experiences and perplexing phenomena on their journey to the Moon. There are storiesof mysterious lights, a lost cosmonaut, a possible extraterrestrial sighting and a photograph of what some believe may be a stargate. See ***Apollo 17*; Grimaldi Crater**.

Schroeter, Johann Hieronymus
(Johann Schroter)
1745–1816

German astronomer and the founder of selenography. Schroeter dedicated his life to the study of the Moon. His observations of the Moon's landscapes were the first to include measurements. On September 26, 1788, Johann Schroeter reported to Johann Bode's *Astronomisches Jahrbuch (Astronomic Yearbook)* that he had witnessed a dazzling white point of light on the Moon. *Reader's Digest's Mysteries of the Unexplained* (Page 238) states that Schroeter observed it shining "to the east of the lunar Alps and in their shadow." It further describes Schroeter 's sighting as "a bright point, as brilliant as a fifth-magnitude star, which disappeared after he had watched it for fifteen minutes." Schroeter also recorded strange changes in the **Linne crater** during his lifetime. He created numerous maps depicting the Moon that included the six-mile long crater over the years. Over time, he observed that Linne was fading. Today Linne is but a small hole with barely any height or depth. Schroeter believed that the Moon was inhabited and that the dwindling of Linne was the result of work by **Selenites**. In an article written by investigative journalist and author Philip Coppens for *Frontier Magazine (1995)* titled "The Alternative Conquest of the Moon," he writes, "As early as 1788 Schroeter had observed small 'swollen parts' on the Moon. He argued that these were the result of industrial activity of the 'Selenites,' the inhabitants of the Moon." See **Schroeter's Valley**.

Schroeter's Valley
(Schroter's Valley; Vallis Schroteri)

A sinuous rille located on the nearside of the Moon. It is named after German astronomer Johann Schroeter. Mysterious activity has been observed in Schroeter's Valley. There has been no explanation for the odd occurrences. On September 16, 1891, American astronomer William Henry Pickering observed something strange in Schroeter's Valley. In NASA's ***Chronological Catalog of Reported Lunar Events***, Pickering's description of what he observed that day is given. He writes, "Dense clouds of white vapour were apparently arising from its bottom and pouring over its SE [IAU:SW] wall in the direction of Herodotus." Pickering also observed what he described as "apparent volcanic activity" on each of the following dates: September 17, 18, 23, 25 and October 14, 1891; May 10, 1892; January 30, 1893; June 14, October 8, 10, 13 and 15, 1897; and April 6, 7, and 8, 1898. Additional accounts of unexplained activity within Schroeter's Valley include an occurrence on February 10, 1949, where an astronomer by the name of Thornton witnessed what he described as a, "Diffuse patch of thin smoke or vapor from W side of Schroeter's Valley near Cobrahead, spreading into plain; detail indistinct, hazy (surrounding area clear)." On April 21, 1967, an unexplained red color was seen by an astronomer named Darnell. Darnell monitored the area for twenty-one hours and twenty minutes. No explanations of any of the recorded anomalous events have been given.

Schroter, Johann
See **Johann Schroeter**.

Scott, David Randolph
1932–

Former **Apollo** astronaut and the seventh person to walk on the Moon. Scott was in the third group of **astronauts** chosen by NASA in October, 1963. His missions included *Gemini 8,* where he served along with **Neil Armstrong**. ***Apollo 9*** where he served as the Command Module pilot and ***Apollo 15*** in which he was the commander. Scott was also an officer in the United States Air Force. During the *Apollo 15* mission, Scott witnessed two inexplicable events on the Moon. In a picture that was taken of Scott performing a drilling operation, an object that resembles a spaceship can be seen hovering above. It appears to be watching Scott as he performs his work on the lunar **surface**. In another episode, Scott and astronaut **James Irwin**, as they again worked on the surface, were nearly struck by a mysterious object that whizzed by. What it was that nearly hit the two men remains unknown. Scott wrote about his trip to the Moon stating, "As I stand out here in the wonders of the unknown at Hadley, I sort of realize there's a fundamental truth to our nature. Man must explore…and this is exploration at its greatest."

Sea of Serenity
See **Mare Serenitatis**.

Sea of Tranquility
See **Mare Tranquillitatis**.

Searching for Alien Artifacts on the Moon

A scientific paper written by Dr. Paul Davies and Robert Wagner that was submitted to the journal, *Acta Astronautica*. The paper proposed SETI expand its search for **extraterrestrial** intelligence beyond its current practice which is listening for **radio signals**. They suggested that extraterrestrials may have already dispatched probes to our portion of the galaxy. See ***Lunar Reconnaissance Orbiter (LRO)***.

Selene

Greek goddess of the Moon. In Greek mythology, Selene drove her chariot across the sky each night. She was often depicted as sitting on a **crescent moon,** or with the crescent Moon placed over her head, or wearing it as a decorative ornament atop her head. She was believed to be the Moon incarnate. Many words are derived from the name Selene including **Selenites** (lunar inhabitants), selenocentric (when speaking of the center of the Moon), selenology (studies involving Moon astronomy), senotropic (to turn towards the Moon), selenotropism (in botany, to turn upwards in response to moonlight). There is also a selenite crystal believed to contain healing properties.

Selenite(s)

A name for lunar inhabitants. The word Selenite is believed to be derived from the name of the Greek goddess Selene. From the time the telescope was created for centuries after, it was considered scientific fact by many that the Moon was an inhabited world. Famed German astronomer Johannes Kepler started the search for the mysterious lunar people often referred to as Selenites in 1610. The famed astronomer **William Herschel** believed that humans are not alone in the universe and spent time searching for **extraterrestrials**. He studied the Moon intensely at night, searching for signs of life. He concluded that there were extraterrestrials living on the Moon, inside the cra**ters**. He stated, "Who can say that it is not extremely probable, nay beyond doubt, that there must be inhabitants on the Moon of some kind or another?" Joseph Smith, the founder of Mormonism, allegedly commented on there being lunar inhabitants. He is quoted from a third party as having said, "The moon is inhabited by men and women the same as on the Earth and they live for a greater age than we do—that they live generally to near the age of 1,000 years." As telescopes became more sophisticated and knowledge of the Moon became more accessible, it became clear that the Moon was uninhabitable. See **Camille Flammarion; Lunarians; Moon Dwellers; Emanuel Swedenborg.**

Selenographer(s)
See **Thomas Gwyn Elger; Hugh Percy Wilkins**.

Shard, The

A large, erect edifice that resembles a **monument**. It stands 1.5 miles above the lunar **surface**. It is located in the Moon's Ukert region. It was famously named "the Shard" by writer, researcher, Richard Hoagland. The picture of the Shard was taken by NASA's ***Lunar Orbiter 3***, which launched on February 5, 1967. The Shard can be viewed on NASA image *AS10-32-4822*.

Shcherbakov, Alexander

Soviet scientist and co-originator of the **spaceship moon theory**, which was first published in an article for *Sputnik* magazine in 1970 titled "Is the Moon a Creation of Alien Intelligence?" There, he and Soviet scientist **Michael Vasin** explained their hypothesis that the Moon is an artificial **satellite** and spacecraft. See **Spaceship Moon Theory**.

Shepard, Alan
1923–1998

Former NASA **astronaut**. Shepard has the distinction of being the first American to travel into space on May 5, 1961. He flew on a Mercury spacecraft. The name of his capsule was Freedom 7. In addition, he was one of NASA's original seven astronauts. He was also the commander of ***Apollo 14*** and one of the few people that have ever walked on the Moon. Shepard went into space, headed towards the Moon, with a secret. Shepard, who was an avid golfer, came up with the idea of hitting a golf ball while on the Moon. He and a friend planned for him to take a special club head that would connect onto an instrument used for collecting samples, and two golf balls that he sneaked onto the craft in a sock. He carried the items to the Moon. Once he reached the lunar **surface**, he took a shot, hitting a ball that was in the air for 35 seconds and reached 200 yards (on Earth it would have been approximately 35 yards). About his experience on the Moon, Shepard has been quoted as saying, "When I first looked back on the Earth, standing on the Moon, I cried."

Shoemaker, Eugene
1927–1998

A famous astronomer and geologist. He founded the field of Planetary Science. It was Shoemaker who created the system and procedures that the Apollo **astronauts** used in studying the Moon. It had been Shoemaker's dream to one day work as an astronaut in space and visit the Moon. Unfortunately, due to a medical problem, Shoemaker was denied entrance into the **Apollo program**. He died in 1998 at the age of 69. Although Shoemaker was unable to fulfill his dream, NASA paid tribute to him by sending his ashes to the Moon in the Lunar Prospector in 1998. The Moon is Shoemaker's final resting place.

Shrinking Moon

The Moon is becoming smaller. Scientists have found cracks in the lunar **surface** which indicate that the Moon is shrinking. First reported by scientists in 2010, the cracks were first discovered in pictures that were taken by NASA's Lunar Reconnaissance Orbiter.

The cracks are due to two phenomena. First, the temperature of the core of the Moon is decreasing, causing some of the internal liquid material to solidify, leading to a reduction in the volume of the Moon. This shrinkage causes cracks on the surface. Second, just as the Moon has a tidal effect on the surface of the Earth, the Earth's gravitational pull has a tidal effect on the **surface** of the Moon. Together with the Sun, they tug on the Moon's solid surface and create faults in the lunar crust.

Signals

Lights have been observed on the lunar **surface** that appear as though someone is signaling Earth. Reports of these intermittent flashing lights date back to the 1800s. On July 4, 1832, British astronomer Thomas Webb witnessed a series of blinking dots and dashes that were reminiscent of Morse code. On October 20, 1824, European Moon observers saw alternating flickering lights, occurring continuously throughout the night, at the Aristarchus crater. In 1873, after carrying out a comprehensive investigation of the flashing lights, the Royal Society of Britain announced that the flashing lights on the Moon were issued by intelligent beings whose goal was to signal Earth.

Silverstein, Abe
1908–2001

Former director of NASA's Space Flight Development Program. It was Silverstein that named the **Apollo program**. He chose the name **Apollo** after being inspired by images of the Greek god in his chariot traversing the skies. See **Apollo; Apollo Missions.**

Sitchin, Zecharia
1920–2010

American author, researcher, expert on the Sumerian culture and ancient **astronaut** proponent. He was the author of the famed *Earth Chronicles Series*, which included *The 12th Planet*, *The Stairway to Heaven*, *The Wars of Gods and Men*, *The Lost Realms*, *When Time Began*, *The Cosmic Code*, *The End of Days: Armageddon and Prophecies of the Return* and *Genesis Revisited*. He was also one of the few people who could read and interpret ancient Sumerian and Akkadian texts. He advanced the Tiamat planetary theory. Sitchin maintained that the Hebrew God was an **extraterrestrial** entity. He traced the origin of the Hebrew religion to the Sumero-Babylonian deities, whom he asserted were the Nephilim (giants) found in the biblical book of Genesis 6:4. He writes that these Nephilim were the gods of the Old Testament, or as they are literally named in the Hebrew texts, the Elohim. He also believes that these extraterrestrial beings originated on a planet called Nibiru per the Sumerians. In his book *Genesis Revisited*, Sitchin wrote about how the Moon came to be, per Sumerian cosmological beliefs. Sitchin deemed that Earth's Moon was originally a satellite for a planet named Tiamat. Tiamat was believed to be located out past Mars. It was a time when the entire solar system had become highly unstable due to developing gravity disturbances. The planet Tiamat had eleven moons. One of those moons increased to a highly abnormal size, which became problematic and even violent for the surrounding planets. The name of this moon was Kingu. Because of this great celestial commotion, Tiamat was broken into two pieces. One piece was demolished, while the other, which was near Kingu, was propelled into a new orbit, becoming Earth. Thus, the Earth and Moon partnership was born.

Size
See **Diameter**.

Skeleton

Chinese astrophysicist Dr. Kang Mao-pang tells a strange tale of a human skeleton that was found on the Moon. The story is believed to have originated in a tabloid paper. Per the tale, the ***Apollo 11*** Lunar Module photographed a human skeleton in 1969. Chinese astrophysicist Dr. Kang Mao-pang released photographs of shoeless human **footprints** located on the Moon. He allegedly was given the strange photographs by US sourccs, asscrting that thc US had conspired to cover up a skeleton being found on the Moon. He is quoted as saying, "They hid photos of bare human footprints on the Moon for 20 years and managed to keep the human skeleton secret even longer."

Purportedly, Mao-pang had numerous pictures that showed the footprints and a human skeleton on the Moon. More shocking information about the skeleton came to light when it was revealed that it was partially dismembered and was wearing jeans. It is believed too that from an examination of the pictures, the person perished violently. There is one theory that the skeleton was taken into space after the person was already dead. The disintegration of the body would not have been possible in the airless **atmosphere** of the Moon. Dr. Mao-pang gave additional thoughts and comments on the finding stating that his source was American and "beyond reproach." He stated that he had top-secret papers confirming that the footprints on the moon were new. The information also stated that the deceased person was undeniably human. One theory of how the skeleton arrived on the Moon is that it was taken there by **extraterrestrials**.

Slayton, Donald K. "Deke"
1924–1993

Former NASA astronaut with the Mercury program. Slayton was one of the original Mercury seven **astronauts** chosen because of his extensive flying experience. The Mercury missions were a prelude to **Project Gemini** and also the **Apollo Program**. Slayton spotted a **UFO** during one of his trips into space. He initially believed it to be a weather balloon until he moved closer and got a better view of it. Slayton stated that the UFO resembled a "saucer sitting on edge, at about a 45 degree angle." He stated that "it took off, climbed at about a 45 degree angle and just accelerated and disappeared." Some wonder if the early astronauts, thosc bcforc thc Apollo missions, were being watched by otherworldly beings due to the future Moon mission agenda, and if early **UFO** incidents (such as Slayton's) might have been related to this.

Soil

The soil covering the lunar **surface** is two to three feet deep. It is comprised mainly of broken rock that has been crushed due to meteorites hitting the Moon over eons of time. See **Orange Soil**.

Sound

The Moon has no atmosphere and, as such, it is soundless. Without an atmosphere to allow sound waves to propagate, it is a completely silent world, just like much of the space outside of the Earth's atmosphere.

Soviet Union

The Soviets had twenty Moon missions and several notable accomplishments. They were the first to take a picture of the **far side** of the Moon and the first to accomplish a Moon landing (with their unmanned *Luna 2* spacecraft on September 13, 1959). Additionally, they had several other firsts in the space race. They had the "first probe to impact the Moon," the "first flyby" the "first soft landing," the "first lunar orbiter," and the "first circumlunar probe to return to Earth." Reportedly, the Soviets took photographs of the lunar floor that reveal constructions that appear to be created with intelligent design. The items include eight large **pyramid** structures. Similarly, there are smaller objects with the same form that are spread out in a configuration comparable to the layout of the three large pyramids of Giza, Egypt. Soviet scientists Mikhail Vasin and Alexander Shcherbakov were proponents of a theory that the Moon is an ancient spaceship, created elsewhere in the universe by superior beings. They wrote an article published in July 1970 in *Sputnik* magazine titled "Is the Moon the Creation of Alien Intelligence?" The two explained their theory that the Moon is really a hollowed out **planetoid**. In the article Vasin and Shcherbakov wrote, "Abandoning the traditional paths of 'common sense,' we have plunged into what may at first sight seem to be unbridled and irresponsible fantasy." They held that these advanced beings used large machinery to create the spaceship. Once they achieved this, the spaceship moon was transported across the universe and placed in Earth's orbit. They maintained that the Moon being a spaceship answered many of the perplexing questions associated with the Moon. Seeing the Moon as a large hollowed out ship that came from elsewhere in the universe would explain why the Moon appears to be significantly older than Earth. It would explain **moonquakes**, the reason behind astronomers witnessing vaporous clouds, the strange results of the seismic testing when the Moon **rang like a bell**, why refractory metals were found in **moon rocks,** and more. These two researchers felt that the spaceship moon theory was the closest that anyone had come to in explaining the existence of the Moon and how it came to be in Earth's trajectory.

Space Music

Apollo 10 **astronauts** heard strange sounds when orbiting the Moon that they referred as "space music." Nearly four decades after the ***Apollo 10*** mission, recordings were found that indicated something strange had happened as *Apollo 10* traveled on the **far side** of the Moon. After the astronauts returned from their trip, the tapes from onboard the spacecraft were transcribed and filed away inside the NASA archives until 2008. These tapes are said to contain weird "music" that came from the radio. The astronauts can be heard commenting on the strange noise on the transcript. Experts have explained the music away as radio interference. It has been argued that the Apollo **astronauts** would know the difference between radio static and interference. Many believe that the sounds were something far more mysterious. See **Appendix 1: NASA Transcripts (*Apollo 10*).**

Space Portal
See **Stargate**.

Spacecraft(s)
See **Spaceships**.

Spaceship(s)

A manned vehicle used for space travel. **UFO** researchers have found what they believe to be remnants of an **extraterrestrial** spacecraft that crashed on the Moon. Researchers also claim that they found footprints leading to the **crashed** object. See **George Adamski; Aristarchus Crater.**

Spaceship Moon Theory

The theory that the Moon is a gigantic spaceship brought here from outside of our solar system. There are some who believe that the Moon is not a natural **satellite**, but was artificially created by **extraterrestrials** and brought into Earth's orbit thousands of years ago. Supporters of this theory not only believe that it is artificial, but that it is a spaceship that is occupied. The idea that the Moon is a spaceship, engineered by advanced extraterrestrials with obvious superior technology, was first introduced by Soviet Scientists, Alexander Shcherbakov and Michael Vasin. Both men belonged to the distinguished Soviet Academy of Sciences. The two published their hypothesis in July 1970 in an article for the Soviet magazine *Sputnik*. It was titled "Is the

Moon a Creation of Alien Intelligence?" They stated that the Moon is a **planetoid** that was hollowed out in the ancient past, somewhere outside of our solar system. They proposed that the extraterrestrials had used large equipment to dissolve stone and create enormous chasms inside the Moon. Magma from the melted rock poured onto the lunar floor. This massive ship was navigated across the universe and ultimately placed within Earth's orbit. In *Sputnik* magazine Shcherbakov and Vasin were quoted as saying, "It is more likely that what we have here is a very ancient spaceship, the interior of which was filled with fuel for the engines, materials and appliances for repair work, navigation instruments, observation equipment and all manner of machinery… everything necessary to enable this 'caravelle of the Universe' to serve as a **Noah's Ark** of intelligence, perhaps even as a home of a whole civilization envisaging a prolonged (thousands of millions of years) existence and long wanderings through space (thousands of millions of miles)." The spaceship moon theory is believed by some researchers to be the one hypothesis that resolves all of the outstanding Moon mysteries. See **Zulu**.

Spherical UFOs
See **UFOs**.

Sphinx

A giant, construction resembling the sphinx of **Egypt** has been located in a NASA photograph, near the **Tycho crater**. Ancient Egypt is believed by some researchers to have a connection to the cosmos. Lunar researchers believe that the figure is evidence of an ancient alien connection between the Moon and the Earth. The framework of the alleged moon sphinx is said to be organized in the same manner as the Sphinx of Egypt. It is estimated to be the same size as the Egyptian sphinx, with the same lines and a similar shape to the head. One noted difference is that the sphinx in the photograph of the Moon is believed to be older. Skeptics believe the formation to simply be a rock.

Spires
See **Blair Cuspids**.

Splitting of the Moon

A miracle in Islamic beliefs that was performed by the Prophet Mohammad. According to the story, Mohammad was receiving opposition to his claim of being a true Prophet of God. His skeptics informed him that if he were truly a prophet, then he should split the Moon into two halves. Mohammad, asked them whether, if he did so, they would believe. They answered yes. Mohammad then pointed to the Moon and through the power of Allah, the Moon was split in two. One half of the Moon could be seen above the mount of Qubais, the other over the mount of Qaiqa'an. Verse 54:1 of the Koran (the Islamic sacred book) states, "The Hour has come near and the moon has split [in two]." The story of the splitting of the Moon has caused some researchers to look for geological evidence of it in the modern era. There are believers that assert that there is evidence of the splitting of the Moon in photographs taken during NASA missions. While many believers maintain that this miracle occurred in the ancient past, there are others that believe that the splitting of the Moon will occur on judgment day.

Spur (Crater)
See **Genesis Rock**.

Stargate

A portal to other worlds. The astronauts of *Apollo 17* took photographs of what lunar researchers believe could be a stargate. NASA picture AS17/AS17-151-23127 shows a mysterious illuminated object on the lunar floor. In an article titled "Did Apollo 17 find a Stargate on the Moon," researcher and author Michael Salla writes, "The object appears to be a space portal of some kind with an eerie blue glowing ring around a central darker portion." Salla offers an in-depth analysis of the picture and the potentiality of this being an actual stargate on ExoNews: "Did Apollo 17 find a Stargate on the Moon?" SO1E13, https://www.youtube.com/watch?v=yd-c2xw1Swc.

Stooke, Dr. Philip
See **Lunar Map at Knowth**.

Stopover

A place where travelers take a break before proceeding to their destination. Some speculate that if **extraterrestrials** were heading toward the Earth, there is a good chance that they would make a stopover on the Moon. Says famed British journalist Harold T. Wilkins in his book *Flying Saucers from the Moon,* "I have suggested that the moon may be and long has been a stopover place for what we call flying saucers, or **spaceships**." *In Flying Saucers on the Attack*, he mentions it again stating, "It was near the close of the 18th century when several men in London and Norwich saw strange lights on the moon that appear to indicate that our **satellite** was being used as a stopover place, in flights of observation to the Earth." Some believe if Wilkins' hypothesis is correct, it may explain the mystery of **UFOs** around Earth's vicinity and the many different types of alleged spaceships that have been reported.

Straight Wall

An atypical formation on the lunar **surface** that measures about seventy mles long. Some lunar researchers believe it to be artificially made. It has been dubbed the "Straight Wall," but is sometimes referred to as the "Railway." As time progressed, astronomers determined that what they were looking at was a large fault line. The fault line is pushing up through the lunar floor, which contributes to its mysterious appearance.

Striped Glasses

An anomalous item that was seen by NASA **astronauts** while near the Moon's Sea of Storms that resembled a pair of reading glasses.

Structure(s)

Respected astronomers, astronauts, researchers and more have been reporting seeing alleged artificial structures on the Moon for centuries. Some come from Moon observations through the telescope, however, most come from images from manned and unmanned missions to the Moon. The types of structures allegedly found include: geometric shaped buildings, crystal structures, platforms, bridges, domes, pyramids, huge spires, large towers and more. Some of the structures are said to resemble artifacts found on Earth. See **Archimedes Platform; Astronauts; Boxed Structure; John Brandenburg; Bridge; Crosses; Dark Side; Preston Dennett; Domes; Extraterrestrial Bases; Extraterrestrials; Glass-Crystal Structure; Industrial Complex; Kepler Crater; *Luna 9*; Madler's Square; Howard Menger; Edgar Mitchell; Monuments; Ruins; Vito Sacchari; Sea of Serenity; Soviet Union; Ingo Swann; Tower of Babel; Tycho Crater; Washington National Press Club Meeting; Karl Wolfe.**

Sun Gate at Tiahuanaco

A large, rock-solid, ancient megalithic stone structure known located in Bolivia. It is roughly three meters tall and is carved from a monolithic block of stone that weighs 10 tons. On it is found mysterious symbols and figures that have astronomical meanings. The Sun God, Viracocha (the god of creation), is located in the middle with sun rays emanating from his head. Other figures bear a likeness to humans, but appear to be a cross between humans and

An early photo of the Sun Gate at Tiahuanaco.

other creatures. Some have wings, while others have curved tails. Still others are wearing what appear to be helmets. Researchers believe that the gateway was once used as a calendar, one that seems to indicate a solar year. However, the solar year depicted does not correspond to the solar year as we divide it today. There is also a belief that the gateway is a portal to another world.

According to the symbols, the Moon arrived in our solar system around 12,000 years ago and in the process caused chaos on Earth. This conclusion was reached in 1956 by researcher and author **Hans Schindler Bellamy** who examined the Sun Gate. He wrote about it in his book *The Calendar of Tiahuanaco*. Bellamy interpreted the ancient astronomical symbols found on the Sun Gate. They tell a story of the history of the Moon. Bellamy explains that the characters on the stone hold mathematical and astronomical information. Besides indicating when the Moon arrived, they state that before the Moon came in, the Earth revolved slower and had only 290 days. Because of this difference in time, it is believed that the Earth was turning on its axis at a more leisurely speed than today and therefore had lengthier days. Bellamy's interpretation of the symbols tells us that these characteristics of Earth changed with the arrival of the Moon. The Moon's arrival would have caused great turmoil on Earth due to the gravitational pull. There would have been an increase in storms, earthquakes and floods and there would have been a great many disasters. It is thought that this event may have been the source of the **great flood** in the tale of **Noah's Ark**, as well as other flood stories of various cultures around the world. The Sun Gate had been placed on a platform temple known as Kalasasaya. Kalasasaya is located next to a second temple that is partially underground. Together, the two temples comprise part of what is believed to be an ancient observatory.

Surface

The surface area of the Moon is 15 million square miles (1/14 that of the Earth). The surface gravity is 5.31 ft/s^2 (1.62 m/s^2, 16.54 % that of the Earth's). The surface temperature is 273 degrees Fahrenheit during the day and minus 280 degrees Fahrenheit at night. The surface dust measures 1/8 inch to 3 inches thick. The makeup of the Moon's surface is a mystery to some researchers due to the materials found there. Some even question how the Moon's composition could have happened naturally. The surface has a type of powdery substance on top of it. Beneath, it is exceptionally hard, with corrosion-resistant titanium being found as one of the elements. **Rocks** brought back by the Apollo **astronauts** were shown to contain processed metals such as brass, mica, uranium 236 and neptunium 237. None of these materials are known to occur organically. Rust resistant iron particles were also discovered. These findings have helped to fuel the idea, that the Moon is a spaceship created to withstand time, space travel and even meteorite strikes.

Swann, Ingo Douglas
1933–2013

Ingo Swann was a prominent American psychic researcher, parapsychologist, author and artist. He was a co-creator and proponent of remote viewing (the ability to see and gather impressions and information about a distant place, location or person through extrasensory perception). Swann was hired by a clandestine agency to view the Moon's far side by means of remote viewing. In his book titled *Penetration: The Question of Extraterrestrial and Human Telepathy,* Swann writes that he saw various structures on the dark side of the Moon and witnessed human-like extraterrestrials there as well. He states that these beings had the ability to breathe on the Moon. He witnessed them working on the Moon and digging. Swann claimed to have seen not only **moon dwellers**, but buildings, structures, factories and roads. He mentions seeing workers that appeared to be robotic putting together what appeared to be a laser mechanism. During his remote viewing session, the extraterrestrials became aware of his presence and went after him. Startled, he stopped the session and theoretically left the Moon. Some of the beings followed him back to Earth, frightening Swann, who stayed hidden for weeks after. Some believe that what Swann witnessed during his remote viewing is directly connected to the reasons that the US stopped going to the Moon.

Swedenborg, Emanuel
1688-1772

Famed Swedish spiritual visionary. Swedenborg journeyed to the Moon during an astral experience (which he performed regularly). He returned to tell of having seen **Moon dwellers**. He incorporated one of his Moon experiences into the book *Life on Other Planets*, where he told of seeing lunar people described as being very short (the size of a seven year old child) with thunderous voices that emanated from their stomach areas using air that accumulates there. It was clear that they were able to exist in an entirely different situation from humans. In his book *Life on Other Planets* (page 87), Swedenborg states, "Spirits and angels know that there are inhabitants on the moon too and equally on the moons or satellites around the planets Jupiter and Saturn. Those who have not seen spirits from there and talked with them are still in no doubt that there are people on those moons, because they are just as much worlds; and where there is a world, man is to be found."

Sword
See **Birt Crater**.

T

Tanit

A sky goddess that ruled over the Moon and stars in ancient Phoenician beliefs. She is represented by a triangular apparatus with straight bars supporting the Moon.

Taurus-Littrow

A lunar valley and the landing site of ***Apollo 17***. It is also the site of a mystery. In 1972, as the **astronauts** of the *Apollo 17* mission explored Taurus-Littrow, they purportedly happened upon an enormous box structure. Reportedly, once the camera revealed what the two astronauts were witnessing, the live feed to the public (who were watching the mission on television) was cut, leaving people stunned. According to NASA, the reason for the interruption was a mishap with the camera that was attached to the *Lunar Rover*. According to the story, the *Lunar Rover* inadvertently took a picture of itself. Some researchers remain skeptical and believe that there is a cover-up of what the astronauts saw that day. See **Box Structure; Walter Cronkite**.

Teczistecatl

An Aztec goddess associated with the Moon. She is represented by the four lunar phases: dark moon, waxing moon, **full moon** and waning moon.

Telsa, Nikola
1856–1943

The famed Serbian-American inventor and mechanical engineer. A futurist, Tesla experimented in transmitting **radio signals** to the Moon. He believed that he had received a response.

Temperature

Because the Moon does not have an atmosphere, the temperature variations are severe and highly dependent on the presence of sunlight or shadow. During the day, in the sunlight, temperatures can reach up to 260° F (127° C). During the night, the Moon **surface** is extremely cold, with temperatures running generally around minus 280° F (minus 173° C).

Terrace(s)

A raised flat platform. There was an odd sighting by the *Apollo 16* **astronauts** of formations that resemble terraces on the Moon. In an ***Apollo 16*** transmission, they discuss them. See **Appendix 1 *Apollo 16*, Terraces.**

Terrestrial Planet

The Moon is sometimes seen as a "terrestrial planet." This is due to the Moon's size and composition which is comparable to the other terrestrial planets which include Earth, Mars, Venus and Mercury. These planets as well as the Moon all have rocky terrain, distinct geological layers, and have either one or two satellites or none at all. Therefore, when the terrestrial planets are being examined, the Moon is sometimes included. The Moon is not officially recognized as a terrestrial or even a dwarf planet because this exclusive category is based on motion.

Tetrad

A **blood moon** occurs when there is a total **lunar eclipse**. Even though the Earth completely blocks the sun's rays from directly reaching the Moon during a total lunar eclipse, the rays undergo a scattering effect as they pass through the Earth's atmosphere. This scattering allows some portion of the rays to be deflected towards the Moon. The rays having gone through the Earth's atmosphere also get filtered by the large dust particles that are present close to the **surface** of the Earth, allowing only the lower frequencies of the sun's light (reddish colors) to pass through. (This is the same filtering effect that causes the sun to look reddish during sunsets.) Consequently, the filtered, scattered rays of the sun' rays reaching the Moon give the impression that the Moon is crimson-colored. Scientists refer to four blood moons in row a as a "tetrad." Even though they are atypical, some centuries occasionally will have a succession of four blood moons.

Theia Impact
See **Giant Impact Theory**.

Theodorus of Cyrene
465 BC–398 BC

Theodorus of Cyrene was a distinguished mathematician of the 5th century BC. Plato was one of his pupils. He was noted for his work in astronomy, music and a number of other scholastic topics. Theodorus once wrote that the Moon had emerged sometime prior to the war in which Hercules battled the giants. This tells us that he believed that the Earth was **moonless** in the beginning and that the orb came into our solar system at a later date, after the Earth was already populated. Some speculate that advanced beings with superior technology brought the Moon into Earth's orbit to aid in the evolution of life on Earth.

Thornton, F.H.
See **Plato Crater**.

Thoth

Egyptian Moon god. In art, Thoth is often depicted with a **crescent moon** that is cradling the disk of the Moon. It was believed that when Ra, the god of the Sun, traveled to the underworld at dusk, Thoth, the wisest of all the gods, ruled the sky until dawn.

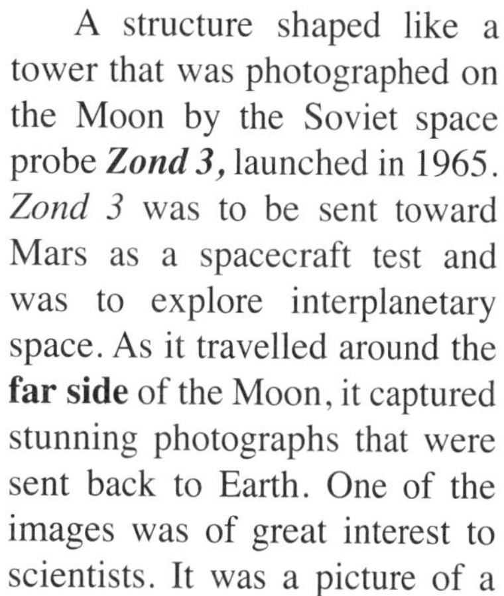

Tiahuanaco

See **Sun Gate at Tiahuanaco**.

Tiamat Planetary Theory

A theory compiled by author and researcher **Zechariah Sitchin,** which tells us that the Moon initially revolved around the planet Tiamat. Per Sitchin's theory, there was a **war** on Tiamat that was so fierce that it caused the planet to be broken into two pieces. One of those pieces was destroyed and is now the asteroid belt. The other part became the Earth. When this occurred, Tiamat's moon, called Kingu at the time, then became Earth's **satellite**.

Tibet(an)

In ancient Tibetan lore, there is a tale of a lost continent named **Gondwana** that had a civilization during a period before there was a Moon in the sky.

Time Before the Moon

See **Pre-Lunar Earth**.

Tower

A curious object photographed by NASA's ***Lunar Orbiter 3*** which launched on February 5, 1967. The object, according to researchers is over a mile thick and approximately five miles high. It is famously known as the "Tower," as named by writer and researcher Richard Hoagland of the "Enterprise Mission." The Tower can be viewed on NASA image AS10-32-4822.

Tower of Babel

A structure shaped like a tower that was photographed on the Moon by the Soviet space probe ***Zond 3,*** launched in 1965. *Zond 3* was to be sent toward Mars as a spacecraft test and was to explore interplanetary space. As it travelled around the **far side** of the Moon, it captured stunning photographs that were sent back to Earth. One of the images was of great interest to scientists. It was a picture of a formation that appeared to be unnatural. It was estimated to be 3.5 miles in height. Because the construction protruded up out of a **crater** and was shaped like a tower, it was dubbed the Tower of Babel. The image puzzled researchers as they tried to explain the unusual object. The same structure can be seen in NASA photographs, including picture AS10-32-4856. There have been many points of view on this particular anomaly. Skeptics believe that the structure is an optical illusion from the sun casting shadows. Some ufologists maintain that it is a launch pad that belongs to **extraterrestrials**. Some speculate that it is a part of an extraterrestrial base, while others believe that it is an artifact from an ancient city that existed long before humans made their journeys to the Moon.

TR-277

See **NASA Technical Report TR-277.**

Tracks

Tracks were one of the unusual items found on the lunar **surface** by the **Apollo** astronauts. How they arrived there is a mystery. During their trip to the Moon's Apennine Mountains, the *Apollo 15* **astronauts** found tracks that could not be explained. According to sources, astronaut Harrison Schmitt the Lunar Module pilot on the ***Apollo 17*** mission, excitedly made the following comment, "I see tracks running right up the wall of the crater." Reportedly this was the response to his sighting. "We copy, Gene. Your photopath runs directly between Pierce and Pease. Pierce Bravo, go to Bravo, whisky, Whiskey, Romeo." **UFO** researchers feel that there was a code in that response so that further discussion of what was being seen was covered up. See **Appendix 1: NASA Transcripts (*Apollo 15*), Tracks**.

Transient Lunar Phenomenon (TLP)
(Lunar Transient Phenomenon, LTP)

A term coined by British astronomer **Patrick Moore**. Transient Lunar Phenomena are irregular, unexplained lights on and around the Moon. They include short-lived lights, orbs of light, **flashes of light**, unexplained illuminations, luminosities and color. Hundreds have been reported from reputable astronomers, scientists and **astronauts**, as well as amateur astronomers and laypeople. Ever since the invention of the telescope in 1608, astronomers have witnessed TLPs on the Moon. Every Apollo astronaut reportedly saw mysterious lights during their mission. TLPs were seen on the Moon so often by astronomers that NASA created a catalog of them titled *Chronological Catalogue of Reported Lunar Events* (also known as ***NASA Technical Report TR-277***). The catalog lists over 570 strange lunar events. In addition, NASA started an investigation titled Operation Moon-Blink, where NASA requested observatories from around the world to monitor and photograph the Moon. Within just a few months of Operation Moon-Blink being created, twenty-eight TLPs were recorded.

Tarot card for the Moon.

Translucent Moon
See **Transparent Moon**.

Transparent Moon

There have been reports from astronomers that when looking at the Moon, they could see all the way through it and see stars. This has led some to consider that the Moon is either translucent or holographic. This effect has been spotted several times during both the waxing and waning Moon.

Trash
See **Man-made Materials**.

Travel Time

Travel time to the Moon by *Apollo 11* was three days. By car it would take 135 days at 70 mph (113 km/h).

Triangular Protuberances

In the book *Strange Universe* by American, physicist and writer William R. Corliss, there is an astonishing account of the Moon having large triangular protuberances first appearing on the top and then disappearing, and later, two more materializing on the bottom and again vanishing. The story was retold in *Reader's Digest's Mysteries of the Unexplained*. According to the account, on the night of July 3, 1882, a group of people who lived in Lebanon, Connecticut observed a peculiar occurrence on the Moon. It was explained that two dark, obelisk-shaped protuberances materialized on the Moon's upper extremity. It is said that with the weird knobs, the Moon resembled a horned owl. A few minutes after they appeared, the protuberances slowly vanished, with the one on the southeast side dissipating first. Minutes later, two other obelisk-shaped protuberances appeared. These were located on the lower portion of the Moon. The notches slowly moved toward each other, next to the rim. It looked as though they were eliminating almost a quarter of the Moon's **surface** as they were moving, until they finally came together. At that point, the Moon suddenly changed back to its natural appearance. According to *Mysteries of the Unexplained*, "When the notches were nearing each other the part of the moon seen between them was in the form of a dove's tail."

Triangular UFOs

For over a century, astronomers have reported seeing glowing, triangular-shaped **UFOs** traveling above the Moon; date as far back as the late 1800s. Sightings of triangular UFOs have also been reported in the skies of Earth for years. They are said to range from approximately 200 to 300 feet and move at extremely high rates of speed. Some UFO proponents have considered the possibility that there is a connection between the triangular UFOs seen near the Moon and those spotted in close proximity to Earth.

Triesnecker (Crater)

A prominent impact **crater** that is named after Austrian Jesuit astronomer Franz de Paula Triesnecker (1745–1817). It is situated close to the Sinus Medii, near the central portion of the nearside of the Moon. Reportedly, a pyramidal formation was found near Triesnecker. On July 10, 1966, a luminous streak of light was seen within the crater. It was observed for one hour.

Tycho (Crater)
(Also Tyco Crater)

A young, distinct impact **crater**. It is considered to be one of the Moon's most mysterious areas. Named after Danish astronomer Tycho Brahe (1546–1601), Tycho is estimated to be around 108 million years old, over 65 miles (104.6 km) across and nearly 3.1 miles (5,000 meters) in depth. It has a mountain in the center that is 1.24 miles (2,000 meters) in height. Tycho can be found in the southern highlands. There is a claim that a giant, **sphinx**-like figure has been located in the Tycho crater. It is estimated to be approximately 230 feet (70 meters) in height and resembles the sphinx of **Egypt**. Ancient Egypt is believed by some researchers to have a connection to the cosmos and **extraterrestrials**. Researchers believe that the alleged structure is evidence of an ancient extraterrestrial connection between the Moon and the Earth. Purportedly, the framework of the sphinx-like structure is organized in the same manner as the one in Egypt. It is estimated to be the same size, with the same lines and a similar shaped head. One noted difference is that the alleged sphinx on the Moon is thought to be older. Skeptics believe the formation to be simply a rock. Best-selling author **Mike Bara** offers a detailed report in his book *Ancient Aliens on the Moon*, about strange, anomalous constructions found on the rim of Tycho, observed in images taken by the ***Clementine*** spacecraft. Among the many mysterious **objects** featured in the book and which Bara has named include: the **Backhoe,** the **Chalet,** the **Geo-dome,** the **Longhorn** and the **Pyramid**. Tycho is also the location of the enigmatic monolith in filmmaker **Stanley Kubrick's** masterpiece ***2001: A Space Odyssey***. A number of unexplained lights and other activity have been seen in and around Tycho. A compilation follows.

1905, August 15, the entire crater shone vividly during a **lunar eclipse**.

1901, April 1, the crater was illuminated during the lunar eclipse.

1919, November 7, in the vicinity of the crater a glowing light could be seen throughout a lunar eclipse.

1919, July 14, a "faint Milky-looking luminosity was seen."

1940, December 9, a brilliant light was seen on the west crater rim of the west outer slope area.

See John **Brandenburg; Moon Base**.

U

Ubell, Earl

See **Rosetta Stone of the Planets**.

UFO(s)

(Unidentified Flying Objects)

For decades, there have been reports of UFOs flying above, across and below the Moon. **Astronauts** on their missions saw UFOs and in some cases, their spacecraft were followed by them. In the 1950s, there allegedly were government tracking stations located underground in Arizona and Nevada. These stations reportedly tracked **UFOs** that were in close proximity to Earth. It is said that the agencies tracked numerous UFOs to the Moon. Because of the many stories surrounding UFOs and the Moon, many have come to believe that the Moon may be occupied. Others reason that the many UFOs seen near the Moon are the result of it being a **stopover** point for otherworldly beings. See **Buzz Aldrin; *Apollo 7; Apollo 8, Apollo 9, Apollo 10; Apollo 11; Apollo 12; Apollo 17;* Neil Armstrong; Astronauts; William R. Brooks; China; City Ship; L. Gordon Cooper Jr.; Cylindrical-Shaped UFOs; Dark Side; Fastwalkers; *Gemini 1; Gemini 9A; Gemini 10;* Germans; Jade Rabbit; Kareeta; Donald Keyhoe; Christopher Kraft; James Lovell; Kenneth Mattingly; James McDivitt; Moon Landing Conspiracy Theory; Moon Pigeons; Objects Crossing the Moon; Plato Crater; Pyramids; Sea of Tranquility; Spaceport; Triangular UFOs; Edward White**.

Ukert (Crater)

An impact **crater** that is named after the German history scholar Friedrich August Ukert. It sits between Mare Vaporum to the north and Sinus Medii to the south. It has a triangular shape and appears to some to be made from intelligent design. There are a number of anomalous objects and constructions near Ukert.

Unidentified Lunar Objects (ULOs)

A term that refers to anomalous objects located on the lunar **surface**.

Urey, Harold

1893–1981

American Nobel Prize-winning scientist. Urey believed that there was a great disproportion of density between the Earth and Moon. This led him to believe that the Moon may be hollow. In *Secrets of Our Spaceship Moon* by Don Wilson, Urey is quoted as saying, "the origin and history of the Moon have remained a mystery despite intensive study by eminent scientists during the last century and a half."

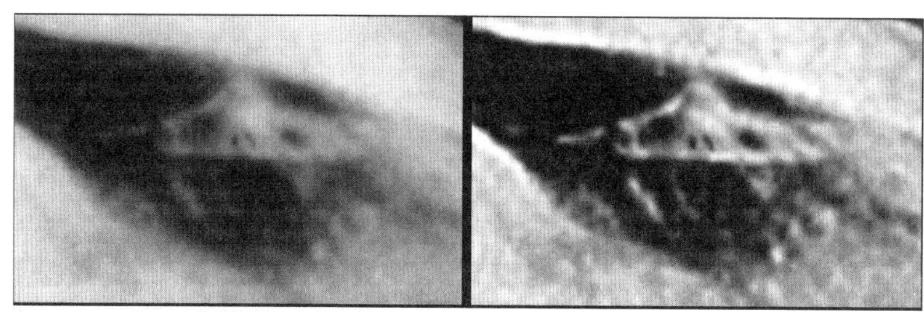

A saucer-shaped UFO parked on the Moon? From *Alien Bases on the Moon.*

Moon a Creation of Alien Intelligence?" The writers of the article were Vasin and Soviet scientist **Alexander Shcherbakov.** In the article the two explained their hypothesis of the Moon being an artificial **satellite** and spacecraft. See **Spaceship Moon Theory.**

V

Vallis Alpes

A stunning lunar valley that some moon researchers believe may have been created by artificial means. It measures 80 miles long and nearly 6 miles wide.

Vallis Schroteri
(Schroeter's Valley; Schroter's Valley)
See **Schroeter's Valley**.

Van der Pol, Balthasar
See **Radio Signals**.

Van Tassel, George Washington
1910–1978

Extraterrestrial **contactee**, researcher, Ufologist and author. His books include: *I Rode a Flying Saucer*; *Into This world and Out Again*; *The Council of Seven Lights*; *Religion and Science Merged* and *When Stars Look Down*. Van Tassel claimed to have personally witnessed antigravity technology of **extraterrestrials** that have been occupying the Moon for eons.

Vasin, Michael
(Unknown birth and death dates)

Soviet scientist and co-originator of the **spaceship moon theory**, first published in an article for *Sputnik* magazine in 1970 titled "Is the

Vegetation

For many centuries it was believed that there was life on the Moon. Many astronomers claimed to have seen plants and foliage and even a type of grass. As telescopes became more sophisticated and more was learned about the Moon, it was concluded that the Moon cannot sustain any type of life. Notwithstanding, speculation about there being vegetation on the Moon continued through the 1950s. See **George Adamski; Alex Collier; Nicolas Camille Flammarion; Franz von Paula Gruithuisen; Howard Menger.**

Velikovsky, Immanuel
1895–1979

American writer and independent scholar. Velikovsky was an exponent of controversial theories as to the origins of the universe. He also questioned and reinterpreted the events of ancient history in his best-selling book *Worlds in Collision*. In his writing, Velikovsky made mention of a time in history when there was no Moon. He quoted from Democritus, **Anaxagoras of Clazomenae**, **Aristotle** and **Apollonius of Rhodes** to prove that such a pre-Hellenic time existed.

Verne, Jules Gabriel
1828–1905

French writer, author, poet and playwright. In 1865, the publication of Verne's novel *From the Earth to the Moon* inspired serious thought about traveling to the Moon.

Villa Jr., Paul
See **Extraterrestrial Bases**.

Vital Statistics
See **Circumference; Diameter; Distance from Earth; Gravity; Length of Day; Orbital Period; Radius; Size; Surface; Temperature; Travel Time; Weight**.

Walker, Joseph (Joe) Albert
1921-1966

Former NASA pilot. Walker was also a captain in the United States Air Force and was a pilot in World War II. His missions included flying X-15 Flight 90 and X-15 Flight 91, in 1963, giving him the distinction of having flown the first two space plane flights. He was the first to fly the Lunar Landing Research Vehicle (used for establishing piloting and operational procedures for future Apollo missions to the Moon). **Neil Armstrong** later flew the same craft to prepare for the historic ***Apollo 11*** mission. One of the responsibilities of Walker on his X-15 flights was to look out for and monitor **UFOs**. On two of his assignments, the film footage taken with his onboard cameras showed strange objects near his craft. They have been variously described as either tubular or disc-shaped. During the *Second National Conference on the Peaceful Uses of Space Research* held in Seattle, Washington, Walker told the audience that he had captured five **cylindrical-shaped UFOs** on film in April 1962, while flying aboard the X-15. He also stated that in March of 1962 he recorded two disc-shaped **UFOs**. One still image from Walker's film was eventually released and can be viewed today.

War

Ancient aliens on the Moon proponents and some lunar researchers believe that there may have been a **war** on the Moon in the ancient past. This theory comes from the conditional that alleged **ruins** and artifacts were found in, and the amount of what appears to be artificial debris located there. Some believe that this war may have been connected to the legendary **Atlantis**. Other's associate it with the Sumerian gods, the **Anunnaki**.

Warned off the Moon

It is said that the reason that the Apollo program ended was because the astronauts were "warned off the Moon" by **extraterrestrials**. These otherworldly beings are believed by some to have encountered by Apollo **astronauts,** and conveyed that the earthlings were not welcome and were not to return. This event is believed by many in the **UFO** community to have during the ***Apollo 11*** mission. Some believe that **Neil Armstrong** and **Buzz Aldrin** were confronted by **extraterrestrials**. It is said that they were told to leave and to never return. This reportedly occurred during the infamous two-minute loss of radio signal, where the astronauts were transmitting to Earth while the public listened but then nothing came from the astronauts for two minutes. After ***Apollo 11***, NASA did send **astronauts** to the Moon for several more missions, although, it has been said that they went back each time for a short time, just long enough to take **samples** from the Moon and leave. The ***Apollo 17*** mission was the last of the NASA trips to the Moon. According to the story, the Moon inhabitants considered the **astronauts** to be trespassers. Conspiracy theorists believe that that is the reason the remaining missions, Apollos 18-20, were cancelled. There is also a story that has been circulated that not only were the astronauts warned off the Moon, but they were escorted off. According to this mysterious tale, there is a photograph with a craft hovering above the Moon's **surface** watching the astronauts.

Warring, Scott C.
See **Face on the Moon**.

Washington National Press Club Meeting

On March 21, 1996, a meeting was held at the Washington National Press Club. An excerpt from the official press release states: "NASA scientists and engineers participating in exploration of Mars and Moon reported results of their discoveries at a briefing at the Washington National Press Club on March 21, 1996. It was announced for the first time that man-caused structures and objects had been discovered on the Moon." Scientists at the function maintained that the constructions were being investigated and results would be released at a later date.

Some highlights of the meeting included:

•The Soviet Union having photographs that could prove the presence of **extraterrestrial** occupation on the Moon.

•Numerous photographs revealing several areas of the lunar **surface** where evidence of extraterrestrial activity was apparent.

•Photographs and film footage taken by NASA **astronauts**.

During the meeting, lunar consultants answered questions as to why the public had not been told of the findings. They were quoted as saying, "It was difficult to foresee reactions of the people hearing the news that extraterrestrial creatures were, or are still on the Moon." In addition, the consultants maintained that the information they had about the Moon was still being investigated. They stated that the results of their analysis would be published at a later date.

Water on the Moon

Water was first discovered on the Moon in 2008 by *Chandrayaan-1*, India's first mission to the moon. See **George Adamski; Atmosphere; Bombing; Alex Collier; Nicolas Camille Flammarion.**

Waterman (Crater)

An impact **crater** situated on the **far side** of the Moon. According to researcher Scott Warring of *UFO Sightings Daily*, a NASA photograph taken by *Apollo 15* allegedly shows a "mothership" inside the Waterman crater. The craft purportedly rests on the rim of the crater, near the Tsiolkovsky crater. Pictures and analysis can be found at *UFO Sightings Daily*, http://www.ufosightingsdaily.com/2011/03/nasa-released-10-km-mothership-in.html.

Weight
See **Mass**.

Wet Moon (Cheshire Moon)

The period when the "horns" (or points) of the **crescent moon** point upward in an angular position, taking on the appearance of a smile. Mysterious lights have been seen at various times over the centuries during a wet moon.

White, Edward
1930–1967

Former NASA astronaut and the pilot for *Gemini 4*. Other careers included serving in the US Air Force, in addition to being a test pilot and aeronautical engineer. He has the distinction of being America's first man to walk in space. White was a part of **Project Gemini,** in preparation for the **Apollo** missions to the Moon. In June 1965, while serving on the ***Gemini 4*** mission, White experienced a mystery. White and his partner **James McDivitt** observed a **UFO** moving quickly towards them. As the object flew first above their craft and then beneath it, the **astronauts** feared a collision and sought evasive maneuvers. McDivitt was able to take videos of the **UFO**. The film revealed a glowing **cylindrical-shaped** object, with a stream of light trailing behind it.

White Rays

A name given to rays found to be emanating from lunar **craters.** They have been referred to as "one of the Moon's greatest mysteries." The rays vary in length and run in different directions. Some lunar researchers believe the rays to be made up of dust-like substances. Some are estimated to be over 1,500 miles long, with most being approximately ten miles wide. They have been seen beaming from the **Aristarchus Crater**, **Kepler Crater**, **Copernicus Crater**, **Tycho Crater,** and others.

Whiteout Project

See **Project Whiteout**.

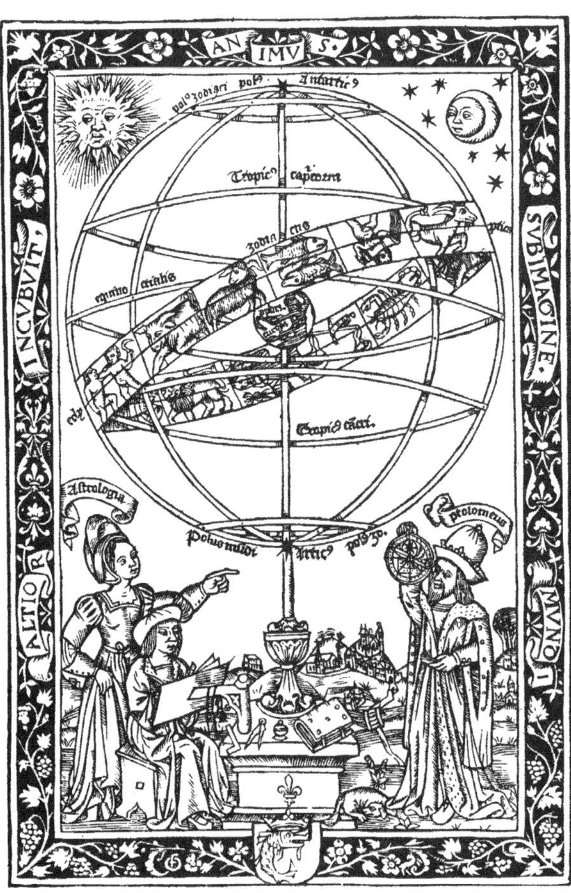

Wilkins, Harold T.

1891–1960

British journalist and author of *Flying Saucers from the Moon and Flying Saucers on the Attack*. Wilkins proposed that the Moon is a stopover point for **extraterrestrials**. After investigating numerous **transient lunar phenomena,** Wilkins wrote in his book *Flying Saucers from the Moon,* "I have suggested that the moon may be and long has been a stopover place for what we call flying saucers, or **spaceships**." *In Flying Saucers on the Attack*, he mentions it again stating, "It was near the close of the 18th century when several men in London and Norwich saw strange lights on the moon that appear to indicate that our **satellite** was being used as a stopover place, in flights of observation to the Earth." Some believe that if Wilkins hypothesis is correct, it may explain the mystery of **UFOs** around Earth's vicinity and the many different types of alleged spaceships that have been reported.

Wilkins, Hugh Percy

1896-1960

A noted British astronomer and selenographer. Wilkins was elected to the British Astronomical Association in 1918 and was the director of the Lunar Section from 1946-1956. He created the popular observers guide, *The Moon: A Complete Description of the Surface of the Moon*, that he co-authored with notable astronomer **Patrick Moore**. He spent years studying the lunar **surface** and created some outstanding drawings charting the **nearside** of the Moon. On the night of March 30, 1950, Wilkins saw an odd glowing object in the Aristarchus-Herodotus area of the Moon. He described it as having an elliptic shape and appeared to be "some type of glowing machine hovering near the crater floor." On August 12, 1944, Wilkins saw what he referred to as "a very bright round spot," near the middle of the **Plato crater**. Wilkins' account is described in the ***Chronological Catalog of Reported Lunar Events***, page 23, number 271. It states, "Exceptional darkness of crater floor, three light spots noted at foot of E wall. Although no light streaks were visible, there was a large and conspicuous spot near the center. Since this spot has been noted as slightly but definitely rimmed all around, Wilkins suggested that temporary dark cloud or vapor may have covered true floor up to level of rim." No explanation was given as to what the spot could have been and it remains a mystery.

Wilkins, John
1614–1672

A dedicated English clergyman, philosopher, amateur scientist and a great thinker and leader of his day. Wilkins contemplated the Moon. He wondered if the Moon is inhabited. He pondered whether men could fly there. He considered the Moon's geography and whether it resembled the Earth in any way. He wrote down his conclusions and published them in an incredible work titled *The Discovery of a World in the Moone* (1638). In his work *A Discourse Concerning a New World and Another Planet* (1640), Wilkins made this prophetic comment: "It is not the bigness of anything in this kind that can hinder its motion, if the motive faculty be answerable thereunto. We see a great ship swims as well as a small cork and an eagle flies in the air as well as a little gnat. 'Tis likely enough that there may be means invented of journeying to the Moon; and how happy they shall be that are first successful in this attempt."

Wilkins, William
(Unknown birth and death dates)

On the evening of March 7, 1794, one William Wilkins reported seeing a light near a darkened area of the Moon. He is quoted in **Harold T. Wilkins'** book *Flying Saucers on the Attack* (1954), describing what he witnessed that evening saying, "This light spot was far distant from the enlightened *(sic)* part of the moon and could be seen with the naked eye. It lasted for 15 minutes and was a fixed and steady light which brightened. It was brighter than any light part of the moon and the moment before it disappeared, the brightness increased. Two persons passing also saw it."

Wolfe, Karl
(Unknown birth and death dates)

Former sergeant in the United States Air Force. In the 1960s Wolf worked as an Air Force Precision Electronics Photographic Repairman. At one point, he was assigned to NASA's Lunar Orbiter program at Langley Air Force Base in Virginia. He was given an assignment to repair machinery in one of NASA's labs. While there he saw personnel processing high-resolution images of the **far side** of the Moon. As he went about his repair duties, Wolfe inquired about the equipment to one of the staff members who was working on the processing of the pictures. The worker confided to Wolf that a **base** had been found on the far side of the Moon. Per Wolfe's report on the matter, the pictures showed an extensive complex that had **structures** in a variety of shapes and sizes, including geometric shapes, towers, mushroom-shaped and spherical constructions.

Worden, Alfred (Al)
1932–

Former NASA **astronaut**, teacher, pilot and the first interplanetary spacewalker. He is the author of *Falling to Earth: An Apollo 15 Astronaut's Journey to the Moon*. Worder served as the Command Module pilot for the ***Apollo 15*** mission to the moon, where he performed scientific experiments, made visual observations, took photographs and mapped the **surface** of the Moon. In recent years, Worden has espoused beliefs in **UFOs** and **extraterrestrials**. In a documentary made to commemorate the twentieth anniversary of the **astronauts'** first landing on the Moon, Worden expressed his beliefs that extraterrestrials had visited Earth in the distant past. Worden stated in an interview that the vision of the biblical prophet Ezekiel was an extraterrestrial spacecraft. He likened it to NASA's Lunar Module. Worden heard a message in an unknown language while in space. According to the story, Worden heard a mysterious transmission come over the radio in what is thought to have been an alien language. He first heard breathing, then a whistling sound and then someone speaking. The words were repeated continually. Linguists were unable to translate it. The odd transmission was recorded and Worden sent it to NASA. The words were, "Mara Rabbi Allardi Dini Endavour Esa Couns Alim."

"X"

The letter X has mysteriously been spotted in the **Eratosthenes crater**. **UFO** proponents maintain that this is a sign that there may have been a base in the area at one time.

Young, John Watts
See ***Apollo 16***.

Zentmayer, Joseph
1826–1888

German American astronomer. **Zentmayer** was famous for making microscopes and other optical devices. In 1869, Zentmayer saw unexplainable objects on the Moon. The objects were illuminated and moving parallel to each other across the Moon during a **lunar eclipse**.

Zetas
(Grays)

An alleged extraterrestrial species from the star system Zeta Reticuli, better known as the **greys**. They have been linked to the many alien **abduction** experiences reported around the world. As a result, there are intricate descriptions of these beings. They are described as having large black eyes with no pupils, large, hairless heads, with just a slit for a mouth. Their ears are described as a small lump on each side of the head. For clothing, they wear what appears to be tight-fitting silver or light blue jumpsuits with a high neck. The Zetas have no lungs, and therefore do not require an atmosphere to exist, and no digestive system. They are said to be emotionless creatures, lacking in spirituality. They are advanced in technology and science and are superior in intellect. Betty and Barney Hill, a couple that claim to have been abducted in the 1960s by the Zetas and whose case has been well documented, were shown a star map by the Zetas in order to show where they were from. Hill later drew a map of Zeta Reticuli, when relaying their abduction experience. There is a theory that the Zetas abduct humans for human DNA, which they are believed to be using in the repairing of their genetic material. Some ufologists believe that the Zetas inhabit the **far side** of the Moon, where they are believed to have a secret installation from which they are observing mankind. These are thought by some to be the beings that warned the **astronauts** to stay away from the Moon. According to Steve Omar in his article "UFOs and Reported Extraterrestrials on the Moon and Mars," the Zetas would be unable to live on Earth's surface due to the atmosphere. They are therefore on the Moon.

Zirna

In Etruscan beliefs, Zirna is the goddess of the waxing moon. Her symbol, which she wears around her neck, is the half moon.

Zond 3

A space probe from the Soviets' Zond program. It was also a part of the Mars *3MV* project. Launched on July 18, 1965, *Zond 3* was a spacecraft test that was to be sent toward Mars. As it executed a flyby of the Moon, it acquired several excellent photographs of the lunar **surface**. One photograph in particular caught the attention of lunar scientists, sparking a mystery. The photograph showed what looked to be a very tall tower that is estimated to be 3.5 miles in height. The picture also showed what appeared to be smaller structures nearby. The tower was named the **Tower of Babel**. Ufologists believe it to be an artificial structure that was made by **extraterrestrials**, while skeptics contend that it is a trick of the light.

Zulu

A Bantu ethnic group of South Africa. Ancient Zulu people referred to the Moon as an "egg." This is because they believed that the Moon had been hollowed out. A Bantu legend states that the Moon was brought into Earth's orbit ages ago by two **extraterrestrial** brothers named Wowane and Mpanku. The brothers were the rulers of the Chitauri (which means **Reptilian**s in Zulu). They were known as the water brothers and had scales over their bodies, similar to fish. They stole the egg (Moon) from the "Great Fire Dragon," and removed the yolk making it hollow. They then proceeded to move the "egg" across the heavens to the Earth, causing great chaos throughout the planet. The account is very similar to modern-day theories of the Moon being hollowed out by extraterrestrials using advanced machinery and then being brought into Earth's orbit, being placed in a perfect position to support life.

Appendix 1: NASA Transcripts

Key

Capcom:	Capsule Communicator
CDR:	Commander
MC:	Mission Control
LMP:	Lunar Module Pilot
Orion:	Lunar Module

Mercury-Atlas 6

John Glenn: This is Friendship Seven. I'll try to describe what I'm in here. I am in a big mass of some very small particles that are brilliantly lit up like they're luminescent. I never saw anything like it. They round a little; they're coming by the capsule and they look like little stars. A whole shower of them coming by. They swirl around the capsule and go in front of the window and they're all brilliantly lighted. They

probably average maybe 7 or 8 feet apart, but I can see them all down below me, also.
Capcom: Roger, Friendship Seven. Can you hear any impact with the capsule? Over.
John Glenn: Negative, negative. They're very slow; they're not going away from me more than maybe 3 or 4 miles per hour. They're going at the same speed I am approximately. They're only very slightly under my speed. Over. They do, they do have a different motion, though, from me because they swirl around the capsule and then depart back the way I am looking.

Gemini 7
Lovell: Bogey at 10 o'clock high.
Capcom: This is Houston. Say again 7.
Lovell: Said we have a bogey at 10 o'clock high.
Capcom: *Gemini 7*, is that the booster or is that an actual sighting?
Lovell: We have several...actual sightings.
Capcom: Estimated distance or size?
Lovell: We also have the booster in sight...

Gemini 10
Gemini 10: This is *10*, Houston. We have two bright objects up here in our orbital path. I don't think they are stars—they look like we are going right along with them.
MC: *10*, Houston.
Gemini 10: *10*, go
MC: Where are the objects from you?
Gemini 10: Roger.
MC: If you can get us a bearing, maybe we can track them down.
Gemini 10: They just disappeared. I guess they were satellites of some kind.

Apollo 8
LMP: There's a strange light down there.
CDR: Is it a bonfire?
LMP: It might be campfires.
CDR: Well, how about that big one out there, in–with the–Is that a, a—very symmetrical pyramid mountain out there? Isn't that Icarus? Huh?
LMP: I would guess that that is Icarus.
CDR: I would guess.
LMP: That is Icarus
CDR: I don't know whether you can see it out your window or not. You see that very pyramid-like central peak?

Apollo 10, Space Music
LMP: That music even sounds outer-spacey, doesn't it? You hear that? That whistling sound?
CDR: Yes.
LMP: Whoooooo.
CMP: Did you hear that whistling sound, too?
LMP: Yes, Sounds like–you know, outer space type music.
CMP: I wonder what it is.
CDR: What the hell was that gurgling noise?
LMP: I don't know. But I'll tell you, that eerie music is what's bothering me.
CMP: God damn, I heard it, too.
LMP:: You know that was funny. That's just like something from outer space, really. Who's going to believe it?
CMP: Nobody. Shall we tell them about it?
LMP: I don't know. We ought to think about it

some.

Apollo 11

CMP: Boy, there must be nothing more desolate than to be inside some of these craters, these conical ones.

CDR: People that live in there probably never get out.

Apollo 15, Tracks

Scott: Arrowhead really runs east to west.

MC: Roger, we copy.

Irwin: Tracks here as we go down slope.

MC: Just follow the tracks, huh?

Irwin: Right we're (garble). We know that's a fairly good run. We're bearing 320, hitting range for 413 ... I can't get over those lineations, that layering on Mt. Hadley.

Scott: I can't either. That's really spectacular.

Irwin: They sure look beautiful.

Scott: Talk about organization!

Irwin: That's the most organized structure I've ever seen!

Scott: It's (garble) so uniform in width.

Irwin: Nothing we've seen before this has shown such uniform thickness from the top of the tracks to the bottom.

Apollo 16, Flying White Objects

Capcom: You talked about something mysterious...

Orion: O.K. Gordy, when we pitched around, I'd like to tell you about something we saw around the L.M. When we were coming about 30 or 40 feet out, there were a lot of objects—white things—flying by. It looked like they were being propelled or ejected, but I'm not convinced of that.

Capcom: We copy that Charlie."

Apollo 16, Blocks

Duke: These devices are unbelievable. I'm not taking a gnomon up there.

Young: O.K., but man, that's going to be a steep bridge to climb.

Duke: You got—YOWEE! Man—John, I tell you this is some sight here. Tony, the blocks in Buster are covered - the bottom is covered with blocks, five meters across. Besides the blocks seem to be in a preferred orientation, northeast to southwest. They go all the way up the wall on those two sides and on the other side you can only barely see the out-cropping at about five percent. Ninety percent of the bottom is covered with blocks that are 50 centimeters and larger.

Capcom: Good show. Sounds like a secondary ...

Duke: Right out here... the blue one that I described from the lunar module window is colored because it is glass coated, but underneath the glass it is crystalline ... the same texture as the **Genesis Rock** ... Dead on my mark.

Young: Mark. It's open.

Duke: I can't believe it!

Young: And I put that beauty in dry!

Capcom: Dover. Dover. We'll start EVA-2 immediately.

Duke: You'd better send a couple more guys up here. They'll have to try (garble).

Capcom: Sounds familiar.

Duke: Boy, I tell you, these EMUs and PLSSs are really super-fantastic!

Apollo 16, Domes, a Structure and Tunnels

Duke: We felt it under our feet. It's a soft spot. Firmer. Where we stand, I tell you one thing. If this place had air, it'd sure be beautiful. It's beautiful with or without air. The scenery up on top of Stone Mountain, you'd have to be there to see this to believe it—those domes are incredible!

MC: O.K., could you take a look at that smoky area there and see what you can see on the face?

Duke: Beyond the domes, the structure goes almost into the ravine that I described and one goes to the top. In the northeast wall of the ravine you can't see the delineation. To the northeast there are tunnels, to the north they are dipping east to about 30 degrees.

Apollo 16, Terraces

Orion: Orion has landed. I can't see how far the (garble) ... this is a blocked field we're in from the south ray—tremendous difference in albedo. I just get the feeling that these **rocks** may have come from somewhere else. Everywhere we saw the ground, which is about the whole sunlit side, you had the same delineation the ***Apollo 15*** photography showed on Hadley, Delta and Hadley Mountains.

Capcom: O.K. Go ahead.

Orion: I'm looking out here at Stone Mountain and it's got. ...it looks like somebody has been out there plowing across the side of it. The beaches—the benches—look like one sort of terrace after another, right up the side. They sort of follow the contour of it right around.

Capcom: Difference in the terraces?

Orion: No, Tony. Not that I could tell from here. These terraces could be raised out of (garble) or something like that.

Apollo 17, The Condorcet Hotel

MC: Go ahead, Ron.

Evans: O.K., Robert, I guess the big thing I want to report from the back side is that I took another look at the—the —cloverleaf in Aitken with the binocs. And that southern dome (garble) to the east.

MC: We copy that, Ron. Is there any difference in the color of the dome and the Mare Aitken there?

Evans: Yes there is... That Condor, Condorsey, or Condorecet or whatever you want to call it there. Condorcet Hotel is the one that has got the diamond shaped fill down in the uh – floor.

MC: Robert. Understand. Condorcet Hotel.

Evans: Condor. Condorset. Alpha. They've either caught a landslide on it or it's got a – and it doesn't look like (garble) in the other side of the wall in the northwest side.

MC: O.K., we copy that northwest wall of Condorcet A.

Evans: The area is oval or elliptical in shape. Of course, the ellipse is toward the top.

[Note: In her book *ET's Are on the Moon & Mars*, C.L. Turnage asks, "Who is the mysterious 'they've' who have caught a landslide?" She alludes to the idea that the astronauts were inadvertently suggesting that the Condorcet Hotel was manufactured by unknown race.]

Appendix 2: Anomalous Lights

1668
Minister Cotton Mather witnessed a strange starlike light between the points of a crescent moon.

1671
Italian astronomer Gian Cassini observed what appeared to be a white cloud floating above the **surface** of the Moon.

1676
In March 1676, Edmund Halley spotted a large unidentified object near the Moon.

1783, 1787, 1821
Sir William Herschel witnessed unusual lights in and around the Moon on several occasions. In July 1821, German Franz Gruithuisen saw brilliant, small flashing lights on the surface of the Moon.

1788
On September 26, 1788, German astronomer Johann Schroeter reported seeing "a whitish bright spot" that was shining in the area around Mare Imbrium.

1794
In 1794, there were two strange lights witnessed by two separate people. One of those people was the famed astronomer William Wilkins.

1800s
In the early 1800s, prominent astronomer John Herschel observed abnormal lights on the moon and travelling over the Moon during a lunar eclipse.

1826
In 1826, an astronomer recorded a sighting of a huge black cloud moving across the Moon's Mare Crisium.

1847
In 1847, there were reports that bright lights had been seen on the Moon during a lunar eclipse.

1867
In 1867, Thomas Gwyn Elger witnessed an odd light suddenly appear on the moon.

1869
In 1869, Joseph Zentmayer observed illuminated objects moving across the Moon during a lunar eclipse.

1871
See **Cecil Maxwell Cade**.

1874
In 1874, Monsieur Lamery reported seeing mysterious black objects moving across the Moon. In the same year, he witnessed a brilliant light traversing the Moon.

1877
In 1877, astronomer Henry Harrison reportedly witnessed an odd light on the Moon. On February 20 of that year, lights were seen near the Eudoxus crater. Also, a Dr. Klein reported seeing a strange triangular light in the Plato crater.

1899
In 1899, a brilliantly lighted object was seen traveling over the Moon's surface. The sighting was reported by two Moon observers, a Dr. Warren E. Day and a Mr. G. Scott both from Arizona.

1915
On April 23, 1915, an unexplained beam of light

was reportedly seen inside the Moon's Clavius crater.

1920 and 1930
During the 1920s and 1930s there were numerous reports of lights traveling across the Moon from lunar observers in the US, Britain and France. Newspapers and journals of the day, published information on the strange light events taking place on the Moon. American astronomers reported witnessing strange lights that were sometimes glowing and at other times flashing on and off.

1944
On August 12, 1944, an astronomer saw a brilliant shining orb in the Plato crater.

1950
In 1950, British astronomer Hugh Percy Wilkins spotted a strange glowing elliptical light in the Aristarchus-Herodotus area of the Moon. Additionally, Wilkins saw a luminous light around the Pluto crater.

1950s and 1960s
During the period of the 1950s and 1960s numerous lights were witnessed on the Moon by astronomers, inside several craters. Some of the lights are said to have glowed while others held the same intensity.

1953
On September 16, 1953, Rudolph Lippert saw a brilliant flash on the Moon.

1958
On November 3, 1958, Russian astronomer Nikolai Kozyrev witnessed what appeared to be red lighting near the Alphonsus crater.

1963
See **James Greenacre**.

1971
See ***Apollo 14***.

Book Sources

13 Things that Don't Make Sense. Michael Brooks. Vintage. New York, NY, 2008.
A Book of Angels. Sophy Burnham. Guideposts. Carmel, NY, 1990.
Above Top Secret. Timothy Good. Quill. New York, NY, 1988.
Above Top Secret: Uncover the Mysteries of the Digital Age. Jim Marrs. The Disinformation, Company, Ltd., New York, NY, 2008.
The American Heritage Dictionary. Boston, MA. Houghton Mifflin Company, 1982.
Ancient Aliens on the Moon. Mike Bara. Adventures Unlimited Press. Kempton, IL, 2012.
Chronological Catalogue of Reported Lunar Events. Barbara M. Middlehurst, Jaylee M. Burley, Patrick Moore, Barbara L. Welther. NASA. Washington, DC, 1968.
Cosmological Ice Ages. Henry Kroll. Trafford Publishing. Victoria, BC, Canada, 2009.
Cosmos of the Soul: A Wake-Up Call for Humanity. Patricia Cori. North Atlantic Books. Berkeley, CA, 2010.
Dark Mission. Richard C. Hoagland and Mike Bara. Feral House. Port Townsend, WA, 2007.
Elder Gods of Antiquity. M. Don Schorn. Ozark Mountain Publishing. Huntsville, AR., 2008.
Encyclopedia of Gods. Michael Jordan. Facts on File. New York, NY 1993.
Encyclopedia of Mystical and Paranormal Experience. Rosemary Ellen Guiley. Harper's. San Francisco, CA, 1994.
ET's Are on the Moon & Mars. C.L. Turnage. Timeless Voyager Press. Santa Barbara, CA, 2013.
Extraterrestrial Archaeology. David Thatcher Childress. Adventures Unlimited Press. Kempton, IL, 2000.
The Extraterrestrial Encyclopedia. D. Darling. Three Rivers Press. New York, NY, 2000.
The Flying Saucer Conspiracy. Major Donald E. Keyhoe. Henry Holt and Company, Inc. New York, NY, 1955.
Flying Saucers on the Attack. Harold T. Wilkins. Ace Star Books. New York, NY, 1967.
For the Moon is Hollow and Aliens Rule the Sky. Rob Shelsky. GKRS Publications. San Bernardino, CA, 2014.
From Outer Space. Howard Menger. New Saucerian Books. Point Pleasant, West Virginia, 1959.
Genesis Revisited. Zecharia Sitchin. Avon Books. New York, NY, 1990.
The Gods of Eden. William Bramley. Avon Books. New York, NY, 1989.
Hidden Realms, Lost Civilizations and Beings from Other World. Jerome Clark. Visible Ink Press. Canton, MI, 2010.
Invader Moon: Who Brought the Moon and Why. Rob Shelsky. Permuted Press LLC. New York, NY, 2016.
Leap of Faith. L. Gordon Cooper. HarperCollins Publishers. New York, NY, 2000.
Legendary Ladies, 50 Goddesses to Empower you and Inspire You. Ann Shen. Chronicle Books. San Francisco, CA, 2018.
Life on Other Planets. Emanuel Swedenborg. The Swedenborg Foundation and The Swedenborg Society. West Chester, PA and London, England, 2006.
The Lost Realms. Zecharia Sitchin. Bear & Company. Rochester, Vermont, 1990.
One Hundred Thousand Years of Man's Unknown History. Robert Charroux. Berkley Medallion Books. New York, NY, 1971.

The Moon and the Planets: A Catalog of Astronomical Anomalies. William Corliss. Sourcebook Project. Glen Arm, Maryland, 1985.
The Moon Watcher's Companion. Donna Henes. Marlowe & Company, New York, NY, 2004.
Moongate. William L. Brian. Future Science Research Publishing Co., Portland, OR, 1982.
Moonscapes. Rosemary Ellen Guiley. Prentice Hall General. New York, NY, 1991.
Mysteries of the Unexplained. The Reader's Digest Association, Inc., Pleasantville, New York/ Montreal, 1982.
New Atlas of the Moon. Serge Brunier. Firefly Books. Ontario, Canada, 2006.
New Lands, Written in 1925. Charles Fort. Boni and Liveright, Inc., New York, NY, 1923.
Nothing in This Book Is True, But It's Exactly How Things Are. Bob Frissell. North Atlantic Books. Berkeley, CA. 1994.
The Occult: The Ultimate Guide for Those Who Would Walk with the Gods. Colin Wilson. Watkins. London, England, 2015.
Other Worlds Than Ours. C. Maxwell Cade, Taplinger Publishing Co., New York, NY, 1966.
Our Ancestors Came from Outer Space. Maurice Chatelain. Doubleday & Company, Inc. New York, NY, 1977.
Our Cosmic Ancestors. Maurice Chatelain. Light Technology, Publishing. Flagstaff, AZ, 1988.
Our Mysterious Spaceship Moon. Don Wilson. Sphere Books Limited. London, England, 1976.
Out-of-Body-Exploring. Preston Dennett. Hampton Roads Publishing Company, Inc., Charlottesville, VA, 2004.
Return to Earth. Buzz Aldrin. Open Road, Integrated Media. New York, NY, 1973.
Secret Influence of the Moon: Alien Origins and Occult Powers. Louis Proud. Inner Traditions International and Bear & Co. Rochester, VT, 2013.
Secrets of Our Spaceship Moon. Don Wilson. Dell Publishing. New York, NY 1979.
The Secret Space Program and Breakaway Civilization. Richard Dolan. Richard Dolan Press. 2016.
The Source. Art Bell. Penguin Group. New York, NY, 2002.
Strange But True: Mysterious and Bizarre People. Thomas Sleman. Barnes & Noble. New York, NY., 1999.
The UFO Book: Encyclopedia of the Extraterrestrial. Jerome Clark. Visible Inc. Press. Detroit MI, 1998.
UFO Magazine UFO Encyclopedia. William J. Birnes. Pocket Books. New York, NY, 2004.
UFO Sightings of 2006-2009. Scott C. Warring. iUniverse. Bloomington, IN, 2010.
UFOs Among the Stars. Timothy Green Beckley. Global Communications. New Brunswick, NJ. 1992.
Walking Through Walls and Other Impossibilities: The Hybrid Agenda. Milton E. Brener. Xlibris Corporation, Bloomington, IN, 2011.
We Discovered Alien Bases on the Moon, II. Fred Steckling. G.A.F. International. Vista, CA, 1981.
Webster's New World Dictionary. Prentice Hall Press, New York, NY, 1970.
Weird Astronomy: Tales of Unusual, Bizarre and Other Hard to Explain Observations. David A.J. Seargent. Springer. New York, NY, 2010.
Who Built the Moon? Christopher Knight and Alan Butler. Watkins Publishing. London, England, 2005.
Witches and Witchcraft. Mary Ellen Guiley. Facts on File. New York, NY. 1999.
Worlds in Collision. Immanuel Velikovsky. Doubleday, New York, NY 1951.

Periodical Sources

Argosy Magazine. Mysterious "Monuments" on the Moon. Volume 371, Number 2, August, 1970.
Proceedings of the National Academy of Sciences. Volume 55, Number 5, May 15, 1966.

Website Sources

A Brighter Moon, http://www.varchive.org/itb/brigmoon.htm
A facelift for the Moon every 81,000 years, http://m.phys.org/news/2016-10-facelift-moon-years.html
A prominent lunar impact crater found in the southeastern region of the Moon, https://www.findagrave.com/memorial/65131128/frank-august-halstead
A Short Biography of James Clarke Greenacre, By Robert O'Connell and Anthony Cook, http://www.the1963aristarchusevents.com/A_Short_Biography_of_James_C__Greenacre_2013-07-28.pdf
A Time Before The Moon–How Did The Moon Get Here?, https://www.youtube.com/watch?v=9z-DYIHDEsE
Abovetopsecret: NASA #2 Moon, http://www.abovetopsecret.com/forum/thread418333/pg2
Ancient Code: Buzz Aldrin, We Were Ordered Away from the Moon, https://www.ancient-code.com/buzz-aldrin-we-were-ordered-away-from-the-moon/
Ancient Code, https://www.ancient-code.com/official-apollo-mission-transcripts-reveal-fascinating-details-about-aliens/
Ancient Origins: The origins of human beings according to ancient Sumerian texts, http://www.ancient-origins.net/human-origins-folklore/origins-human-beings-according-ancient-sumerian-texts-0065
Alex Collier Moon and Mars Lecture 1996, http://www.alexcollier.org/3
Alien City on Moon Seen in NASA Photo, UFO Researchers Say, http://www.inquisitr.com/2484287/nasa-alien-city-on-moon-ufo/
Alien Conspiracy, A Different Point of View, Exposes Endymion Lunar Operations Command Post, http://alienconspiracy.club/endymion-lunar-operations-command-post/
Alien Contactee and Alien Abductee Cases are Much More Common, http://freedom-articles.toolsforfreedom.com/top-20-alien-contactee-abductee-part-1/
Aliens, Everything you Want to Know, Aliens on the Moon, http://www.aliens-everything-you-want-to-know.com/AliensOnTheMoon.html
Aliens Have Been on the Moon Since Ancient Times, http://ufodigest.com/news/0708/ancient-moon2.html
Aliens on the Moon, http://sillykhan.com/tag/man-on-the-moon/
Aliens on the Moon, http://www.aliens-everything-you-want-to-know.com/AliensOnTheMoon.html
Aliens on the Moon, http://www.nbcnews.com/science/space/aliens-moon-tv-show-adds-weird-ufo-twists-apollo-tales-n159806

Aliens Shot Down NASA LCROSS Probe Before Impact, http://www.bibliotecapleyades.net/luna/esp_luna_54.htm#Aliens_Shot_Down_NASA_LCROSS_Probe_Before_Impact
The Alternative Conquest of the Moon, by Philip Coppens, https://www.bibliotecapleyades.net/luna/esp_luna_13.htm
The Alternative Conquest of the Moon, by Philip Coppens, https://www.eyeofthepsychic.com/moon/
American Museum of Natural History, https://www.amnh.org/exhibitions/beyond-planet-earth-the future of space exploration/rcturning-to-thc-moon/moon-vital-statse Moon takes 29.5 days to orbit the Earth
An Alien Colony Banned to Humans?, http://www.salvation-of-humans.com/English/05-01_e_t_and-lies.htm#colonie_interdite
Anaxagoras, http://www.iep.utm.edu/anaxagor/
Ancient, Shattered Lunar Domes, http://www.bibliotecapleyades.net/luna/esp_luna_79.htm
Ancient Thebit and Huygens's Sword, http://www.skyandtelescope.com/observing/celestial-objects-to-watch/ancient-thebit-and-huygenss-sword/
Angelismarriti.it, *An Alien Spaceship on the Moon: Interview with William Rutledge, Member of the Apollo 20 Crew*, by Luca Scantamburlo, http://www.angelismarriti.it/ANGELISMARRITI-ENG/REPORTS_ARTICLES/Apollo20-InterviewWithWilliamRutledge.htm
Another Interesting Leak, A Second NASA Scientist Tells Us That 'Somebody Else' Is On the Moon, http://www.collective-evolution.com/2016/01/02/another-interesting-leak-a-second-nasa-scientist-tells-us-that-somebody-else-is-on-the-moon/
Apollo 10, https://history.nasa.gov/SP-4029/Apollo_10a_Summary.htm
Apollo 10 Mission Objective: http://www.nasa.gov/mission_pages/apollo/missions/apollo10.html#.VLbhTihBrgE
Apollo 14 astronaut claims peace-loving aliens prevented 'nuclear war' on Earth, http://www.foxnews.com/science/2015/08/14/apollo-14-astronaut-claims-peace-loving-aliens-prevented-nuclear-war-on-earth.html
Apollo 20, Journey into Darkness, https://theunredacted.com/apollo-20-journey-into-darkness/
Apollo Moon Conversations and Pictures show NASA Cover-Up, http://www.bibliotecapleyades.net/luna/esp_luna_4a.htm#Another%20strange%20Apollo%2016%20ground-to-air%
Apollo Moon Conversations and Pictures Show NASA Cover-up, http://www.ufos-aliens.co.uk/cosmicphotos.html
Apollo Space Program, https://www.youtube.com/watch?v=rK1M87sd7eg
The Archimedes Platform, http://www.astrosurf.com/lunascan/luna1.html
The Aristarchus Anomaly: A Beacon on the Moon, http://mysteriousuniverse.org/2013/11/the-aristarchus-anomaly-a-beacon-on-the-moon/
Ashtar-about-the-Moon, http://www.garuda.co/2009/10/27/ashtar-about-the-moon-op-101009/
Astonishing Intelligent Artifacts Found on Mysterious Far Side of the Moon, https://www.bibliotecapleyades.net/luna/esp_luna_35.htm
Astounding Moon Footage, http://www.bibliotecapleyades.net/luna/esp_luna_5.htm
Astronaut John Young, One of Only 12 to Walk on Moon, Dies at 87, http://fortune.com/2018/01/06/astronaut-john-young-dead/
Astronaut "UFO" Sightings, http://debunker.com/texts/astronaut_ufo.html
Astronauts Who Told The World We Are Being Visited & What They Said, http://exonews.org/astronauts-who-told-the-world-we-are-being-visited-what-they-said/

Astronomers Admit to Seeing Triangle UFO's Circling Moon, https://www.top10ufo.com/astronomers-admit-to-seeing-triangle-ufos-circling-moon/
Astronomers Admit to Triangles Transversing the Moon, http://worldufophotosandnews.org/?p=9905
Astronomy, http://www.astronomy.com/columnists/stephen%20omeara/2010/05/stephen%20james%20omearas%20secret%20sky%20oneills%20illusion
Atlantipedia, A-Z Guide to the Search for Plato's Atlantis, http://atlantipedia.ie/samples/blumrich-josef-f/
Atlantipedia, A-Z Guide to the Search for Plato's Atlantis, http://atlantipedia.ie/samples/hoerbiger-hans/
AZ Quotes, https://www.azquotes.com/quote/1283526
Bad Archaeology, https://badarchaeology.wordpress.com/tag/kalasasaya/
Before the Flood, there was no Moon, http://www.pravdareport.com/news/russia/10-10-2002/13385-0/
Before the Moon Existed, https://verumetinventa.wordpress.com/2015/03/08/articles-before-the-moon-existed-part-5-by-raymond-towers/
Before the Moon, Historic Observations, https://verumetinventa2.wordpress.com/2017/06/16/before-the-moon-02-historic-observations-by-raymond-towers/
Berkeley Study of Lunar Cratering History, http://www.bibliotecapleyades.net/ciencia/ciencia_nemesis08.htm
Best Top News, http://www.besttopnews.com/kurs/13-02-2006/17349-princ-0/
The Bible, Genesis 2:7, New International Version, (NIV), https://www.biblegateway.

Author Biography

Constance Victoria Briggs is a metaphysical, spiritual and cosmic researcher and writer. She has authored three books: *The Encyclopedia of Angels*, *Encyclopedia of God*, and *The Encyclopedia of the Unseen World*. Briggs has also been a guest speaker on several radio shows discussing the paranormal, extraterrestrials, life-after-death, near-death-experiences, as well as other related topics. It is Briggs's goal to investigate the mysteries of the universe and how they connect to humanity. She and her family make their home in Southern California.

You can email her at: starseedmission.com

Get these fascinating books from your nearest bookstore or directly from: Adventures Unlimited Press

www.adventuresunlimitedpress.com

ANCIENT ALIENS ON MARS
By Mike Bara

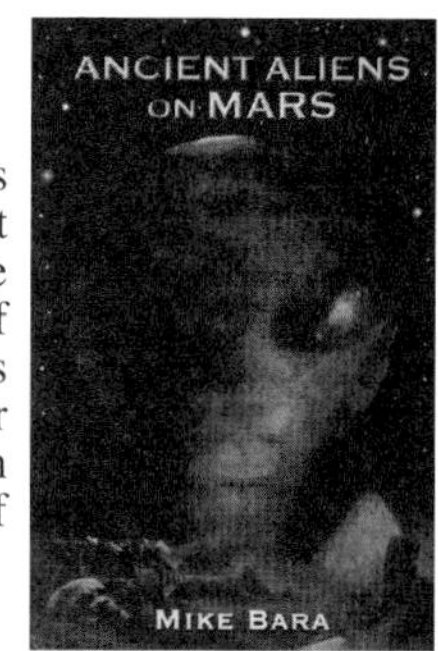

Bara brings us this lavishly illustrated volume on alien structures on Mars. Was there once a vast, technologically advanced civilization on Mars, and did it leave evidence of its existence behind for humans to find eons later? Did these advanced extraterrestrial visitors vanish in a solar system wide cataclysm of their own making, only to make their way to Earth and start anew? Was Mars once as lush and green as the Earth, and teeming with life? Chapters include: War of the Worlds; The Mars Tidal Model; The Death of Mars; Cydonia and the Face on Mars; The Monuments of Mars; The Search for Life on Mars; The True Colors of Mars and The Pathfinder Sphinx; more. Color section.

252 Pages. 6x9 Paperback. Illustrated. $19.95. Code: AMAR

ANCIENT ALIENS ON THE MOON
By Mike Bara

What did NASA find in their explorations of the solar system that they may have kept from the general public? How ancient really are these ruins on the Moon? Using official NASA and Russian photos of the Moon, Bara looks at vast cityscapes and domes in the Sinus Medii region as well as glass domes in the Crisium region. Bara also takes a detailed look at the mission of Apollo 17 and the case that this was a salvage mission, primarily concerned with investigating an opening into a massive hexagonal ruin near the landing site. Chapters include: The History of Lunar Anomalies; The Early 20th Century; Sinus Medii; To the Moon Alice!; Mare Crisium; Yes, Virginia, We Really Went to the Moon; Apollo 17; more. Tons of photos of the Moon examined for possible structures and other anomalies. 8-Page Color Section.

248 Pages. 6x9 Paperback. Illustrated. $19.95. Code: AAOM

PRODIGAL GENIUS
The Life of Nikola Tesla
by John J. O'Neill

This special edition of O'Neill's book has many rare photographs of Tesla and his most advanced inventions. Tesla's eccentric personality gives his life story a strange romantic quality. He made his first million before he was forty, yet gave up his royalties in a gesture of friendship, and died almost in poverty. Tesla could see an invention in 3-D, from every angle, within his mind, before it was built; how he refused to accept the Nobel Prize; his friendships with Mark Twain, George Westinghouse and competition with Thomas Edison. Tesla is revealed as a figure of genius whose influence on the world reaches into the far future. Deluxe, illustrated edition.

408 pages. 6x9 Paperback. Illustrated. Bibliography. $18.95. Code: PRG

THE CRYSTAL SKULLS
Astonishing Portals to Man's Past
by David Hatcher Childress and Stephen S. Mehler

Childress introduces the technology and lore of crystals, and then plunges into the turbulent times of the Mexican Revolution form the backdrop for the rollicking adventures of Ambrose Bierce, the renowned journalist who went missing in the jungles in 1913, and F.A. Mitchell-Hedges, the notorious adventurer who emerged from the jungles with the most famous of the crystal skulls. Mehler shares his extensive knowledge of and experience with crystal skulls. Having been involved in the field since the 1980s, he has personally examined many of the most influential skulls, and has worked with the leaders in crystal skull research, including the inimitable Nick Nocerino, who developed a meticulous methodology for the purpose of examining the skulls.

294 pages. 6x9 Paperback. Illustrated. Bibliography. $18.95. Code: CRSK

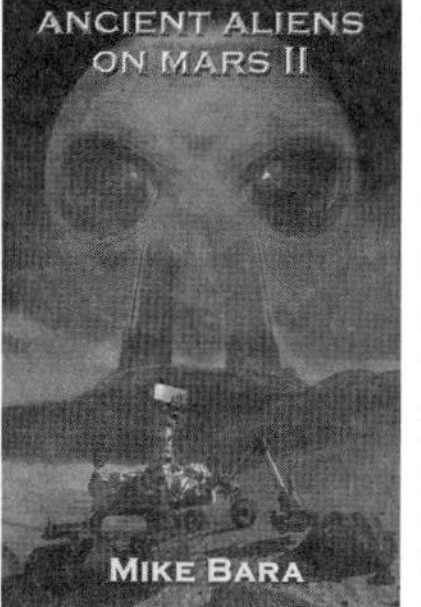

ANCIENT ALIENS ON MARS II
By Mike Bara

Using data acquired from sophisticated new scientific instruments like the Mars Odyssey THEMIS infrared imager, Bara shows that the region of Cydonia overlays a vast underground city full of enormous structures and devices that may still be operating. He peels back the layers of mystery to show images of tunnel systems, temples and ruins, and exposes the sophisticated NASA conspiracy designed to hide them. Bara also tackles the enigma of Mars' hollowed out moon Phobos, and exposes evidence that it is artificial. Long-held myths about Mars, including claims that it is protected by a sophisticated UFO defense system, are examined. Data from the Mars rovers Spirit, Opportunity and Curiosity are examined; everything from fossilized plants to mechanical debris is exposed in images taken directly from NASA's own archives.

294 Pages. 6x9 Paperback. Illustrated. $19.95. Code: AAM2

THE ANTI-GRAVITY FILES
A Compilation of Patents and Reports
Edited by David Hatcher Childress

In the tradition of *The Anti-Gravity Handbook* and *the Time-Travel Handbook* comes this compilation of material on anti-gravity, free energy, flying saucers and Tesla technology. With plenty of technical drawings and explanations, this book reveals suppressed technology that will change the world in ways we can only dream of. Chapters include: A Brief History of Anti-Gravity Patents; The Motionless Electromagnet Generator Patent; Mercury Anti-Gravity Gyros; The Tesla Pyramid Engine; Anti-Gravity Propulsion Dynamics; The Machines in Flight; More Anti-Gravity Patents; Death Rays Anyone?; The Unified Field Theory of Gravity; and tons more. The book that finally blows the lid on suppressed technology, zero-point energy and anti-gravity! Heavily illustrated. 4-page color section.

216 pages. 8x10 Paperback. Illustrated. References. $22.00. Code: AGF

PROJECT MK-ULTRA AND MIND CONTROL TECHNOLOGY
A Compilation of Patents and Reports
By Axel Balthazar

People from around the world claim to be victims of mind control technology. Medical professionals are quick to marginalize these targeted individuals and diagnose them with mental illness. Unfortunately, most people are oblivious to the historical precedent and patented technology that exists on the subject. This book is a compilation of the government's documentation on MK-Ultra, the CIA's mind control experimentation on unwitting human subjects, as well as over 150 patents pertaining to artificial telepathy (voice-to-skull technology), behavior modification through radio frequencies, directed energy weapons, electronic monitoring, implantable nanotechnology, brain wave manipulation, nervous system manipulation, neuroweapons, psychological warfare, subliminal messaging, and more.

384 pages. 7x10 Paperback. Illustrated. References. $19.95. Code: PMK

HIDDEN AGENDA
NASA and the Secret Space Program
By Mike Bara

Bara looks into the Army Ballistic Missile Agency's (ABMA) study to determine the feasibility of constructing a scientific/military base on the Moon. On June 8, 1959, a group at the ABMA produced for the US Department of the Army a report entitled Project Horizon, a "Study for the Establishment of a Lunar Military Outpost." The permanent outpost was predicted to cost $6 billion and was to become operational in December 1966 with twelve soldiers stationed at the Moon base. Did this happen? Did NASA and the Pentagon expect to find evidence of alien bases on the Moon? Did the Apollo 12 astronauts deliberately damage the TV cameras in order to hide their explorations of one of these bases? Does hacker Gary Mackinnon's discovery of defense department documents identifying "non-terrestrial officers" serving in space mean that the US has secret space platforms designed to fight a war with an alien race? Includes an 8-page color section.

346 Pages. 6x9 Paperback. Illustrated. Bibliography. $19.95. Code: HDAG

HAARP: The Ultimate Weapon of the Conspiracy
By Jerry Smith

The HAARP project in Alaska is one of the most controversial projects ever undertaken by the U.S. Government. Jerry Smith gives us the history of the HAARP project and explains how works, in technically correct yet easy to understand language. At best, HAARP is science out-of-control; at worst, HAARP could be the most dangerous device ever created, a futuristic technology that is everything from super-beam weapon to world-wide mind control device. Topics include Over-the-Horizon Radar and HAARP, Mind Control, ELF and HAARP, The Telsa Connection, The Russian Woodpecker, GWEN & HAARP, Earth Penetrating Tomography, Weather Modification, Secret Science of the Conspiracy, more. Includes the complete 1987 Eastlund patent for his pulsed super-weapon that he claims was stolen by the HAARP Project.

256 pages. 6x9 Paperback. Illustrated. Bibliography. $14.95. Code: HARP

TECHNOLOGY OF THE GODS
The Incredible Sciences of the Ancients
by David Hatcher Childress

Popular *Lost Cities* author David Hatcher Childress takes us into the amazing world of ancient technology, from computers in antiquity to the "flying machines of the gods." Childress looks at the technology that was allegedly used in Atlantis and the theory that the Great Pyramid of Egypt was originally a gigantic power station. He examines tales of ancient flight and the technology that it involved; how the ancients used electricity; megalithic building techniques; the use of crystal lenses and the fire from the gods; evidence of various high tech weapons in the past, including atomic weapons; ancient metallurgy and heavy machinery; the role of modern inventors such as Nikola Tesla in bringing ancient technology back into modern use; impossible artifacts; and more.

356 PAGES. 6x9 PAPERBACK. ILLUSTRATED. BIBLIOGRAPHY. $16.95. CODE: TGOD

DEATH ON MARS
The Discovery of a Planetary Nuclear Massacre
By John E. Brandenburg, Ph.D.

New proof of a nuclear catastrophe on Mars! In an epic story of discovery, strong evidence is presented for a dead civilization on Mars and the shocking reason for its demise: an ancient planetary-scale nuclear massacre leaving isotopic traces of vast explosions that endure to our present age. The story told by a wide range of Mars data is now clear. Mars was once Earth-like in climate, with an ocean and rivers, and for a long period became home to both plant and animal life, including a humanoid civilization. Then, for unfathomable reasons, a massive thermo-nuclear explosion ravaged the centers of the Martian civilization and destroyed the biosphere of the planet. But the story does not end there. This tragedy may explain Fermi's Paradox, the fact that the cosmos, seemingly so fertile and with so many planets suitable for life, is as silent as a graveyard.

278 Pages. 6x9 Paperback. Illustrated. Bibliography. Color Section. $19.95. Code: DOM

BEYOND EINSTEIN'S UNIFIED FIELD
Gravity and Electro-Magnetism Redefined
By John Brandenburg, Ph.D.

Brandenburg reveals the GEM Unification Theory that proves the mathematical and physical interrelation of the forces of gravity and electromagnetism! Brandenburg describes control of space-time geometry through electromagnetism, and states that faster-than-light travel will be possible in the future. Anti-gravity through electromagnetism is possible, which upholds the basic "flying saucer" design utilizing "The Tesla Vortex." Chapters include: Squaring the Circle, Einstein's Final Triumph; A Book of Numbers and Forms; Kepler, Newton and the Sun King; Magnus and Electra; Atoms of Light; Einstein's Glory, Relativity; The Aurora; Tesla's Vortex and the Cliffs of Zeno; The Hidden 5th Dimension; The GEM Unification Theory; Anti-Gravity and Human Flight; The New GEM Cosmos; more. Includes an 8-page color section.

312 Pages. 6x9 Paperback. Illustrated. $18.95. Code: BEUF

THE COSMIC WAR
Interplanetary Warfare, Modern Physics, and Ancient Texts
By Joseph P. Farrell

There is ample evidence across our solar system of catastrophic events. The asteroid belt may be the remains of an exploded planet! The known planets are scarred from incredible impacts, and teeter in their orbits due to causes heretofore inadequately explained. Included: The history of the Exploded Planet hypothesis, and what mechanism can actually explode a planet. The role of plasma cosmology, plasma physics and scalar physics. The ancient texts telling of such destructions: from Sumeria (Tiamat's destruction by Marduk), Egypt (Edfu and the Mars connections), Greece (Saturn's role in the War of the Titans) and the ancient Americas.

436 Pages. 6x9 Paperback. Illustrated.. $18.95. Code: COSW

THE GRID OF THE GODS
The Aftermath of the Cosmic War & the Physics of the Pyramid Peoples
By Joseph P. Farrell with Scott D. de Hart

Farrell looks at Ashlars and Engineering; Anomalies at the Temples of Angkor; The Ancient Prime Meridian: Giza; Transmitters, Nazis and Geomancy; the Lithium-7 Mystery; Nazi Transmitters and the Earth Grid; The Master Plan of a Hidden Elite; Moving and Immoveable Stones; Uncountable Stones and Stones of the Giants and Gods; Gateway Traditions; The Grid and the Ancient Elite; Finding the Center of the Land; The Ancient Catastrophe, the Very High Civilization, and the Post-Catastrophe Elite; Tiahuanaco and the Puma Punkhu Paradox: Ancient Machining; The Black Brotherhood and Blood Sacrifices; The Gears of Giza: the Center of the Machine; Alchemical Cosmology and Quantum Mechanics in Stone; tons more.

436 Pages. 6x9 Paperback. Illustrated. $19.95. Code: GOG

ANCIENT ALIENS & SECRET SOCIETIES
By Mike Bara

Did ancient "visitors"—of extraterrestrial origin—come to Earth long, long ago and fashion man in their own image? Were the science and secrets that they taught the ancients intended to be a guide for all humanity to the present era? Bara establishes the reality of the catastrophe that jolted the human race, and traces the history of secret societies from the priesthood of Amun in Egypt to the Templars in Jerusalem and the Scottish Rite Freemasons. Bara also reveals the true origins of NASA and exposes the bizarre triad of secret societies in control of that agency since its inception. Chapters include: Out of the Ashes; From the Sky Down; Ancient Aliens?; The Dawn of the Secret Societies; The Fractures of Time; Into the 20th Century; The Wink of an Eye; more.

288 Pages. 6x9 Paperback. Illustrated. $19.95. Code: AASS

THE ANTI-GRAVITY HANDBOOK

edited by David Hatcher Childress, with Nikola Tesla, T.B. Paulicki, Bruce Cathie, Albert Einstein and others

The new expanded compilation of material on Anti-Gravity, Free Energy, Flying Saucer Propulsion, UFOs, Suppressed Technology, NASA Cover-ups and more. Highly illustrated with patents, technical illustrations and photos. This revised and expanded edition has more material, including photos of Area 51, Nevada, the government's secret testing facility. This classic on weird science is back in a 90s format!

- **How to build a flying saucer.**
- **Arthur C. Clarke on Anti-Gravity.**
- **Crystals and their role in levitation.**
- **Secret government research and development.**

230 PAGES. 7x10 PAPERBACK. ILLUSTRATED. $16.95. CODE: AGH

ANTI–GRAVITY & THE WORLD GRID

Is the earth surrounded by an intricate electromagnetic grid network offering free energy? This compilation of material on ley lines and world power points contains chapters on the geography, mathematics, and light harmonics of the earth grid. Learn the purpose of ley lines and ancient megalithic structures located on the grid. Discover how the grid made the Philadelphia Experiment possible. Explore the Coral Castle and many other mysteries, including acoustic levitation, Tesla Shields and scalar wave weaponry. Browse through the section on anti-gravity patents, and research resources.

274 PAGES. 7x10 PAPERBACK. ILLUSTRATED. $14.95. CODE: AGW

ANTI–GRAVITY & THE UNIFIED FIELD

edited by David Hatcher Childress

Is Einstein's Unified Field Theory the answer to all of our energy problems? Explored in this compilation of material is how gravity, electricity and magnetism manifest from a unified field around us. Why artificial gravity is possible; secrets of UFO propulsion; free energy; Nikola Tesla and anti-gravity airships of the 20s and 30s; flying saucers as superconducting whirls of plasma; anti-mass generators; vortex propulsion; suppressed technology; government cover-ups; gravitational pulse drive; spacecraft & more.

240 PAGES. 7x10 PAPERBACK. ILLUSTRATED. $14.95. CODE: AGU

THE MYSTERY OF THE OLMECS

by David Hatcher Childress

Lost Cities author Childress takes us deep into Mexico and Central America in search of the mysterious Olmecs, North America's early, advanced civilization. The Olmecs, now sometimes called Proto-Mayans, were not acknowledged to have existed as a civilization until an international archeological meeting in Mexico City in 1942. At this time, the megalithic statues, large structures, ceramics and other artifacts were acknowledged to come from this hitherto unknown culture that pre-dated all other cultures of Central America. But who were the Olmecs? Where did they come from? What happened to them? How sophisticated was their culture? How far back in time did it go? Why are many Olmec statues and figurines seemingly of foreign peoples such as Africans, Europeans and Chinese? Is there a link with Atlantis? In this heavily illustrated book, join Childress in search of the lost cites of the Olmecs!

432 Pages. 6x9 Paperback. Illustrated. Bibliography. $20.00. Code: MOLM

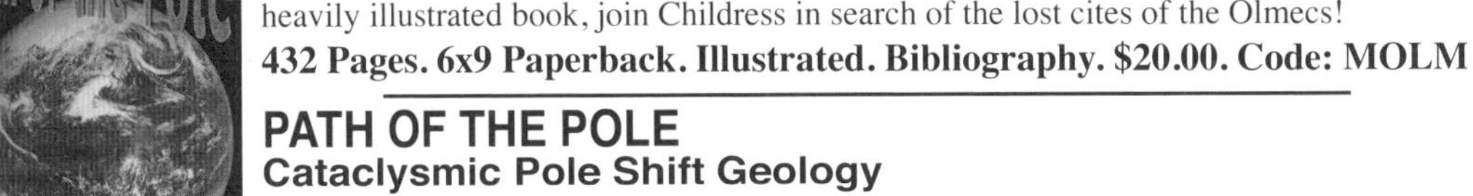

PATH OF THE POLE

Cataclysmic Pole Shift Geology

by Charles H. Hapgood

Maps of the Ancient Sea Kings author Hapgood's classic book *Path of the Pole* is back in print! Hapgood researched Antarctica, ancient maps and the geological record to conclude that the Earth's crust has slipped on the inner core many times in the past, changing the position of the pole. *Path of the Pole* discusses the various "pole shifts" in Earth's past, giving evidence for each one, and moves on to possible future pole shifts. Packed with illustrations, this is the sourcebook for many other books on cataclysms and pole shifts.

356 PAGES. 6x9 PAPERBACK. ILLUSTRATED. $16.95. CODE: POP

MAPS OF THE ANCIENT SEA KINGS

Evidence of Advanced Civilization in the Ice Age

by Charles H. Hapgood

Charles Hapgood's classic 1966 book on ancient maps produces concrete evidence of an advanced world-wide civilization existing many thousands of years before ancient Egypt. He has found the evidence in the Piri Reis Map that shows Antarctica, the Hadji Ahmed map, the Oronteus Finaeus and other amazing maps. Hapgood concluded that these maps were made from more ancient maps from the various ancient archives around the world, now lost. Not only were these unknown people more advanced in mapmaking than any people prior to the 18th century, it appears they mapped all the continents. The Americas were mapped thousands of years before Columbus. Antarctica was mapped when its coasts were free of ice!

316 PAGES. 7x10 PAPERBACK. ILLUSTRATED. BIBLIOGRAPHY & INDEX. $19.95. CODE: MASK

ANCIENT TECHNOLOGY IN PERU & BOLIVIA
By David Hatcher Childress

Childress speculates on the existence of a sunken city in Lake Titicaca and reveals new evidence that the Sumerians may have arrived in South America 4,000 years ago. He demonstrates that the use of "keystone cuts" with metal clamps poured into them to secure megalithic construction was an advanced technology used all over the world, from the Andes to Egypt, Greece and Southeast Asia. He maintains that only power tools could have made the intricate articulation and drill holes found in extremely hard granite and basalt blocks in Bolivia and Peru, and that the megalith builders had to have had advanced methods for moving and stacking gigantic blocks of stone, some weighing over 100 tons.

340 Pages. 6x9 Paperback. Illustrated.. $19.95 Code: ATP

THE ENIGMA OF CRANIAL DEFORMATION
Elongated Skulls of the Ancients
By David Hatcher Childress and Brien Foerster

In a book filled with over a hundred astonishing photos and a color photo section, Childress and Foerster take us to Peru, Bolivia, Egypt, Malta, China, Mexico and other places in search of strange elongated skulls and other cranial deformation. The puzzle of why diverse ancient people—even on remote Pacific Islands—would use head-binding to create elongated heads is mystifying. Where did they even get this idea? Did some people naturally look this way—with long narrow heads? Were they some alien race? Were they an elite race that roamed the entire planet? Why do anthropologists rarely talk about cranial deformation and know so little about it?

250 Pages. 6x9 Paperback. Illustrated. $19.95. Code: ECD

ROSWELL AND THE REICH
By Joseph P. Farrell

Farrell here delves ever deeper into the activities of this nefarious group. In his previous works, Farrell has clearly demonstrated that the Nazis were clandestinely developing new and amazing technologies toward the end of WWII, and that the key scientists involved in these experiments were exported to the Allied countries at the end of the conflict, mainly the United States, in a move called Operation Paperclip. Now, Farrell has meticulously reviewed the best-known Roswell research from UFO-ET advocates and skeptics alike, as well as some little-known source material, and comes to a radically different scenario of what happened in Roswell, New Mexico in July 1947, and why the US military has continued to cover it up to this day. Farrell presents a fascinating case that what crashed may have been representative of an independent postwar Nazi power—an extraterritorial Reich monitoring its old enemy, America, and the continuing development of the very technologies confiscated from Germany at the end of the War.

540 pages. 6x9 Paperback. Illustrated. $19.95. Code: RWR

VIMANA:
Flying Machines of the Ancients
by David Hatcher Childress

According to early Sanskrit texts the ancients had several types of airships called vimanas. Like aircraft of today, vimanas were used to fly through the air from city to city; to conduct aerial surveys of uncharted lands; and as delivery vehicles for awesome weapons. David Hatcher Childress, popular *Lost Cities* author and star of the History Channel's long-running show Ancient Aliens, takes us on an astounding investigation into tales of ancient flying machines. In his new book, packed with photos and diagrams, he consults ancient texts and modern stories and presents astonishing evidence that aircraft, similar to the ones we use today, were used thousands of years ago in India, Sumeria, China and other countries. Includes a 24-page color section.

408 Pages. 6x9 Paperback. Illustrated. $22.95. Code: VMA

BIG BROTHER TECHNOLOGY
PRISM, XKeyscore, and other Spy Tools of the Global Surveillance State
By Axel Balthazar

The government can hack into any computer or smartphone on the planet. What sounded like a crazy conspiracy theory was exposed as truth with the 2013 NSA leaks from Edward Snowden. Since then, the deluge of CIA and NSA hacking programs filling the sky like rain hasn't stopped. This is an exposé of the software programs and techniques used by the agencies to spy on the planet. Big Brother is watching. It's time to watch back. Dozens of previously classified government surveillance programs are divulged in this alarming book! Contents include these fascinating topics: Edward Snowden; NSA; Mass Surveillance; Five Eyes; FinCEN (Financial Crimes Enforcement Network); Stuxnet; PRISM; MYSTIC; DCSNet (Digital Collection System Network); XKeyscore; DISHFIRE; STONEGHOST; Magic Lantern; ECHELON; Fairview; WikiLeaks; Vault 7; Julian Assange; Room 641A; The Doughnut; Fort Meade; Menwith Hill; Utah Data Center; ICREACH; Ransomware, Tor; "wannacry"; ShadowBrokers; and tons more. Axel Balthazar is at it again!

264 pages. 6x9 Paperback. Illustrated. References. $19.95. Code: BBT

HITLER'S SUPPRESSED AND STILL-SECRET WEAPONS, SCIENCE AND TECHNOLOGY
by Henry Stevens

In the closing months of WWII the Allies assembled mind-blowing intelligence reports of supermetals, electric guns, and ray weapons able to stop the engines of Allied aircraft—in addition to feared x-ray and laser weaponry. Chapters include: The Kammler Group; German Flying Disc Update; The Electromagnetic Vampire; Liquid Air; Synthetic Blood; German Free Energy Research; German Atomic Tests; The Fuel-Air Bomb; Supermetals; Red Mercury; Means to Stop Engines; more.

335 Pages. 6x9 Paperback. Illustrated. $19.95. Code: HSSW

SECRETS OF THE MYSTERIOUS VALLEY
by Christopher O'Brien

No other region in North America features the variety and intensity of unusual phenomena found in the world's largest alpine valley, the San Luis Valley of Colorado and New Mexico. Since 1989, Christopher O'Brien has documented thousands of high-strange accounts that report UFOs, ghosts, crypto-creatures, cattle mutilations, skinwalkers and sorcerers, along with portal areas, secret underground bases and covert military activity. This mysterious region at the top of North America has a higher incidence of UFO reports than any other area of the continent and is the publicized birthplace of the "cattle mutilation" mystery. Hundreds of animals have been found strangely slain during waves of anomalous aerial craft sightings. Is the government directly involved? Are there underground bases here?

460 pages. 6x9 Paperback. Illustrated. Bibliography. $19.95. Code: SOMV

QUEST FOR ZERO-POINT ENERGY
Engineering Principles for "Free Energy"
by Moray B. King

King expands, with diagrams, on how free energy and anti-gravity are possible. The theories of zero point energy maintain there are tremendous fluctuations of electrical field energy embedded within the fabric of space. King explains the following topics: Tapping the Zero-Point Energy as an Energy Source; Fundamentals of a Zero-Point Energy Technology; Vacuum Energy Vortices; The Super Tube; Charge Clusters: The Basis of Zero-Point Energy Inventions; Vortex Filaments, Torsion Fields and the Zero-Point Energy; Transforming the Planet with a Zero-Point Energy Experiment; Dual Vortex Forms: The Key to a Large Zero-Point Energy Coherence. Packed with diagrams, patents and photos. With power shortages now a daily reality in many parts of the world, this book offers a fresh approach very rarely mentioned in the mainstream media.

224 pages. 6x9 Paperback. Illustrated. $14.95. code: QZPE

TAPPING THE ZERO POINT ENERGY
Free Energy & Anti-Gravity in Today's Physics
by Moray B. King

King explains how free energy and anti-gravity are possible. The theories of the zero point energy maintain there are tremendous fluctuations of electrical field energy imbedded within the fabric of space. This book tells how, in the 1930s, inventor T. Henry Moray could produce a fifty kilowatt "free energy" machine; how an electrified plasma vortex creates anti-gravity; how the Pons/Fleischmann "cold fusion" experiment could produce tremendous heat without fusion; and how certain experiments might produce a gravitational anomaly.

180 pages. 5x8 Paperback. Illustrated. $12.95. Code: TAP

THE FREE-ENERGY DEVICE HANDBOOK
A Compilation of Patents and Reports
by David Hatcher Childress

A large-format compilation of various patents, papers, descriptions and diagrams concerning free-energy devices and systems. *The Free-Energy Device Handbook* is a visual tool for experimenters and researchers into magnetic motors and other "over-unity" devices. With chapters on the Adams Motor, the Hans Coler Generator, cold fusion, superconductors, "N" machines, space-energy generators, Nikola Tesla, T. Townsend Brown, and the latest in free-energy devices. Packed with photos, technical diagrams, patents and fascinating information, this book belongs on every science shelf. With energy and profit being a major political reason for fighting various wars, free-energy devices, if ever allowed to be mass distributed to consumers, could change the world! Get your copy now before the Department of Energy bans this book!

292 pages. 8x10 Paperback. Illustrated. Bibliography. $16.95. Code: FEH

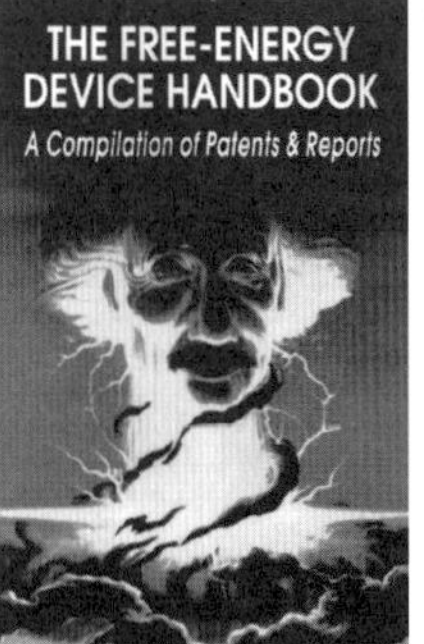

EYE OF THE PHOENIX
Mysterious Visions and Secrets of the American Southwest
by Gary David

GaryDavid explores enigmas and anomalies in the vast American Southwest. Contents includes: The Great Pyramids of Arizona; Meteor Crater—Arizona's First Bonanza?; Chaco Canyon—Ancient City of the Dog Star; Phoenix—Masonic Metropolis in the Valley of the Sun; Along the 33rd Parallel—A Global Mystery Circle; The Flying Shields of the Hopi Katsinam; Is the Starchild a Hopi God?; The Ant People of Orion—Ancient Star Beings of the Hopi; Serpent Knights of the Round Temple; The Nagas—Origin of the Hopi Snake Clan?; The Tau (or T-shaped) Cross—Hopi/Maya/Egyptian Connections; The Hopi Stone Tablets of Techqua Ikachi; The Four Arms of Destiny—Swastikas in the Hopi World of the End Times; and more.

348 pages. 6x9 Paperback. Illustrated. Bibliography. $16.95. Code: EOPX

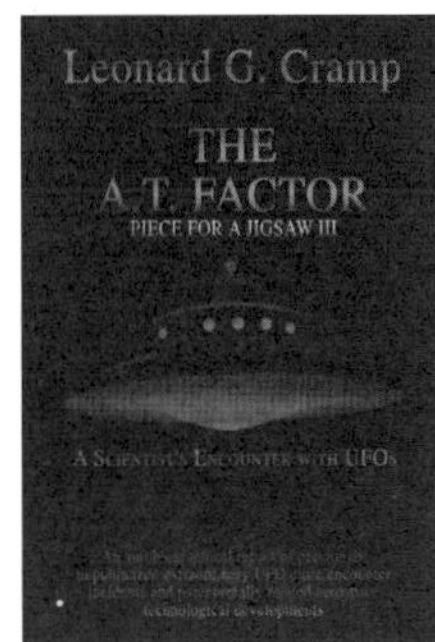

THE A.T. FACTOR

A Scientists Encounter with UFOs: Piece For A Jigsaw Part 3

by Leonard Cramp

British aerospace engineer Cramp began much of the scientific anti-gravity and UFO propulsion analysis back in 1955 with his landmark book *Space, Gravity & the Flying Saucer* (out-of-print and rare). His next books (available from Adventures Unlimited) *UFOs & Anti-Gravity: Piece for a Jig-Saw* and *The Cosmic Matrix: Piece for a Jig-Saw Part 2* began Cramp's in depth look into gravity control, free-energy, and the interlocking web of energy that pervades the universe. In this final book, Cramp brings to a close his detailed and controversial study of UFOs and Anti-Gravity.

324 PAGES. 6X9 PAPERBACK. ILLUSTRATED. BIBLIOGRAPHY. INDEX. $16.95. CODE: ATF

COSMIC MATRIX

Piece for a Jig-Saw, Part Two

by Leonard G. Cramp

Leonard G. Cramp, a British aerospace engineer, wrote his first book *Space Gravity and the Flying Saucer* in 1954. Cosmic Matrix is the long-awaited sequel to his 1966 book *UFOs & Anti-Gravity: Piece for a Jig-Saw*. Cramp has had a long history of examining UFO phenomena and has concluded that UFOs use the highest possible aeronautic science to move in the way they do. Cramp examines anti-gravity effects and theorizes that this super-science used by the craft—described in detail in the book—can lift mankind into a new level of technology, transportation and understanding of the universe. The book takes a close look at gravity control, time travel, and the interlocking web of energy between all planets in our solar system with Leonard's unique technical diagrams. A fantastic voyage into the present and future!

364 PAGES. 6X9 PAPERBACK. ILLUSTRATED. BIBLIOGRAPHY. $16.00. CODE: CMX

UFOS AND ANTI-GRAVITY

Piece For A Jig-Saw

by Leonard G. Cramp

Leonard G. Cramp's 1966 classic book on flying saucer propulsion and suppressed technology is a highly technical look at the UFO phenomena by a trained scientist. Cramp first introduces the idea of 'anti-gravity' and introduces us to the various theories of gravitation. He then examines the technology necessary to build a flying saucer and examines in great detail the technical aspects of such a craft. Cramp's book is a wealth of material and diagrams on flying saucers, anti-gravity, suppressed technology, G-fields and UFOs. Chapters include Crossroads of Aerodymanics, Aerodynamic Saucers, Limitations of Rocketry, Gravitation and the Ether, Gravitational Spaceships, G-Field Lift Effects, The Bi-Field Theory, VTOL and Hovercraft, Analysis of UFO photos, more.

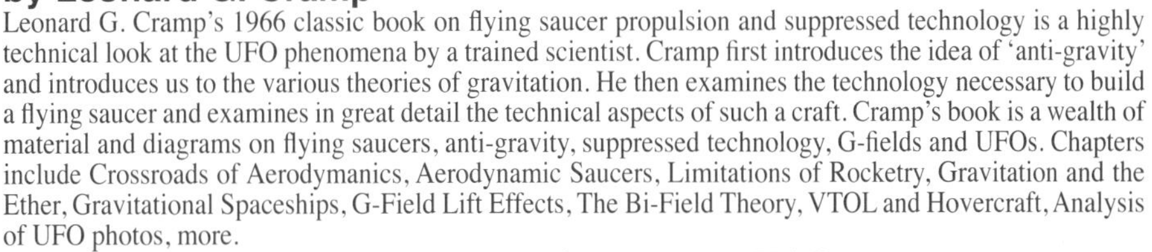

388 PAGES. 6X9 PAPERBACK. ILLUSTRATED. $19.95. CODE: UAG

THE TESLA PAPERS

Nikola Tesla on Free Energy & Wireless Transmission of Power

by Nikola Tesla, edited by David Hatcher Childress

David Hatcher Childress takes us into the incredible world of Nikola Tesla and his amazing inventions. Tesla's rare article "The Problem of Increasing Human Energy with Special Reference to the Harnessing of the Sun's Energy" is included. This lengthy article was originally published in the June 1900 issue of *The Century Illustrated Monthly Magazine* and it was the outline for Tesla's master blueprint for the world. Tesla's fantastic vision of the future, including wireless power, anti-gravity, free energy and highly advanced solar power. Also included are some of the papers, patents and material collected on Tesla at the Colorado Springs Tesla Symposiums, including papers on: •The Secret History of Wireless Transmission •Tesla and the Magnifying Transmitter •Design and Construction of a Half-Wave Tesla Coil •Electrostatics: A Key to Free Energy •Progress in Zero-Point Energy Research •Electromagnetic Energy from Antennas to Atoms •Tesla's Particle Beam Technology •Fundamental Excitatory Modes of the Earth-Ionosphere Cavity

325 PAGES. 8X10 PAPERBACK. ILLUSTRATED. $16.95. CODE: TTP

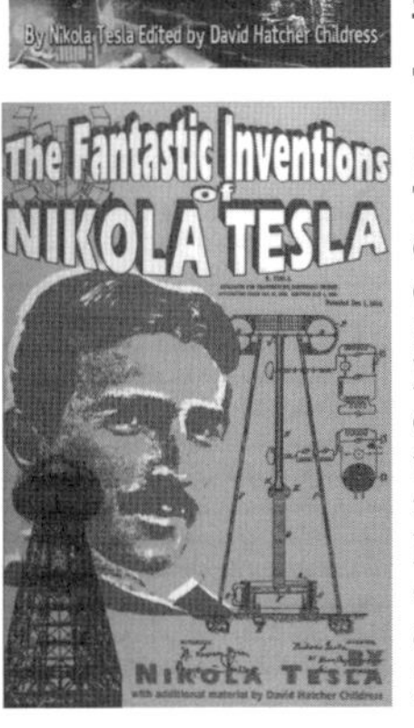

THE FANTASTIC INVENTIONS OF NIKOLA TESLA

by Nikola Tesla with additional material by David Hatcher Childress

This book is a readable compendium of patents, diagrams, photos and explanations of the many incredible inventions of the originator of the modern era of electrification. In Tesla's own words are such topics as wireless transmission of power, death rays, and radio-controlled airships. In addition, rare material on German bases in Antarctica and South America, and a secret city built at a remote jungle site in South America by one of Tesla's students, Guglielmo Marconi. Marconi's secret group claims to have built flying saucers in the 1940s and to have gone to Mars in the early 1950s! Incredible photos of these Tesla craft are included. The Ancient Atlantean system of broadcasting energy through a grid system of obelisks and pyramids is discussed, and a fascinating concept comes out of one chapter: that Egyptian engineers had to wear protective metal head-shields while in these power plants, hence the Egyptian Pharoah's head covering as well as the Face on Mars! •His plan to transmit free electricity into the atmosphere. •How electrical devices would work using only small antennas. •Why unlimited power could be utilized anywhere on earth. •How radio and radar technology can be used as death-ray weapons in Star Wars.

342 PAGES. 6X9 PAPERBACK. ILLUSTRATED. $16.95. CODE: FINT

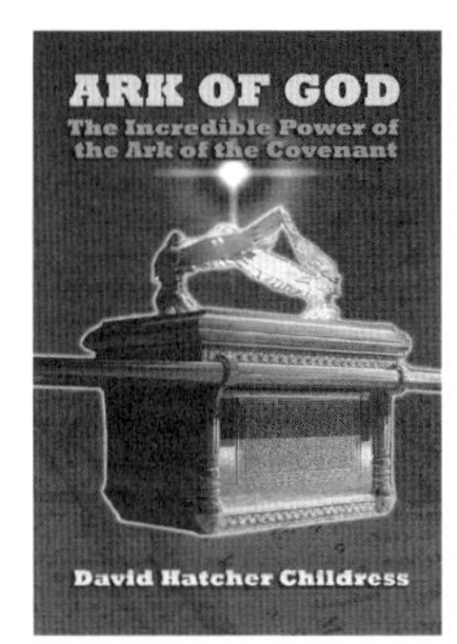

ARK OF GOD
The Incredible Power of the Ark of the Covenant
By David Hatcher Childress

Childress takes us on an incredible journey in search of the truth about (and science behind) the fantastic biblical artifact known as the Ark of the Covenant. This object made by Moses at Mount Sinai—part wooden-metal box and part golden statue—had the power to create "lightning" to kill people, and also to fly and lead people through the wilderness. The Ark of the Covenant suddenly disappears from the Bible record and what happened to it is not mentioned. Was it hidden in the underground passages of King Solomon's temple and later discovered by the Knights Templar? Was it taken through Egypt to Ethiopia as many Coptic Christians believe? Childress looks into hidden history, astonishing ancient technology, and a 3,000-year-old mystery that continues to fascinate millions of people today. Color section.

420 Pages. 6x9 Paperback. Illustrated. $22.00 Code: AOG

SAUCERS, SWASTIKAS AND PSYOPS
A History of a Breakaway Civilization
By Joseph P. Farrell

Farrell discusses SS Commando Otto Skorzeny; George Adamski; the alleged Hannebu and Vril craft of the Third Reich; The Strange Case of Dr. Hermann Oberth; Nazis in the US and their connections to "UFO contactees"; The Memes—an idea or behavior spread from person to person within a culture—are Implants. Chapters include: The Nov. 20, 1952 Contact: The Memes are Implants; The Interplanetary Federation of Brotherhood; Adamski's Technological Descriptions and Another ET Message: The Danger of Weaponized Gravity; Adamski's Retro-Looking Saucers, and the Nazi Saucer Myth; Dr. Oberth's 1968 Statements on UFOs and Extraterrestrials; more.

272 Pages. 6x9 Paperback. Illustrated. $19.95. Code: SSPY

LBJ AND THE CONSPIRACY TO KILL KENNEDY
By Joseph P. Farrell

Farrell says that a coalescence of interests in the military industrial complex, the CIA, and Lyndon Baines Johnson's powerful and corrupt political machine in Texas led to the events culminating in the assassination of JFK. Chapters include: Oswald, the FBI, and the CIA: Hoover's Concern of a Second Oswald; Oswald and the Anti-Castro Cubans; The Mafia; Hoover, Johnson, and the Mob; The FBI, the Secret Service, Hoover, and Johnson; The CIA and "Murder Incorporated"; Ruby's Bizarre Behavior; The French Connection and Permindex; Big Oil; The Dead Witnesses: Guy Bannister, Jr., Mary Pinchot Meyer, Rose Cheramie, Dorothy Killgallen, Congressman Hale Boggs; LBJ and the Planning of the Texas Trip; LBJ: A Study in Character, Connections, and Cabals; LBJ and the Aftermath: Accessory After the Fact; The Requirements of Coups D'État; more.

342 Pages. 6x9 Paperback. $19.95 Code: LCKK

HESS AND THE PENGUINS
The Holocaust, Antarctica and the Strange Case of Rudolf Hess
By Joseph P. Farrell

Farrell looks at Hess' mission to make peace with Britain and get rid of Hitler—even a plot to fly Hitler to Britain for capture! How much did Göring and Hitler know of Rudolf Hess' subversive plot, and what happened to Hess? Why was a doppleganger put in Spandau Prison and then "suicided"? Did the British use an early form of mind control on Hess' double? John Foster Dulles of the OSS and CIA suspected as much. Farrell also uncovers the strange death of Admiral Richard Byrd's son in 1988, about the same time of the death of Hess.

288 Pages. 6x9 Paperback. Illustrated. $19.95. Code: HAPG

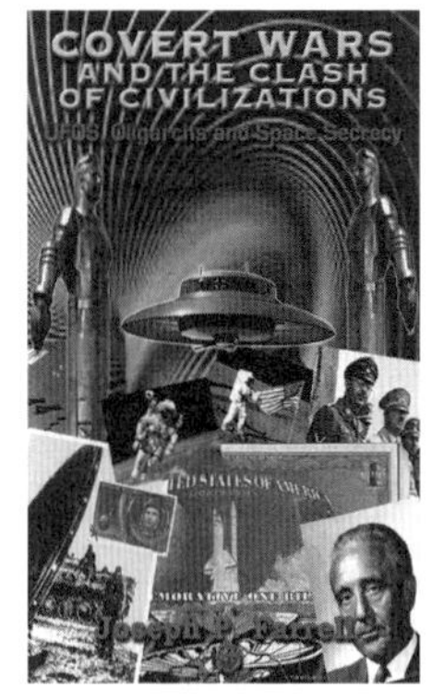

COVERT WARS & THE CLASH OF CIVILIZATIONS
UFOs, Oligarchs and Space Secrecy
By Joseph P. Farrell

Farrell's customary meticulous research and sharp analysis blow the lid off of a worldwide web of nefarious financial and technological control that very few people even suspect exists. He elaborates on the advanced technology that they took with them at the "end" of World War II and shows how the breakaway civilizations have created a huge system of hidden finance with the involvement of various banks and financial institutions around the world. He investigates the current space secrecy that involves UFOs, suppressed technologies and the hidden oligarchs who control planet earth for their own gain and profit.

358 Pages. 6x9 Paperback. Illustrated. $19.95. Code: CWCC

HAARP
The Ultimate Weapon of the Conspiracy
by Jerry Smith

The HAARP project in Alaska is one of the most controversial projects ever undertaken by the U.S. Government. Jerry Smith gives us the history of the HAARP project and explains how works, in technically correct yet easy to understand language. At at worst, HAARP could be the most dangerous device ever created, a futuristic technology that is everything from super-beam weapon to world-wide mind control device. Topics include Over-the-Horizon Radar and HAARP, Mind Control, ELF and HAARP, The Telsa Connection, The Russian Woodpecker, GWEN & HAARP, Earth Penetrating Tomography, Weather Modification, Secret Science of the Conspiracy, more. Includes the complete 1987 Eastlund patent for his pulsed super-weapon that he claims was stolen by the HAARP Project.

256 pages. 6x9 Paperback. Illustrated. Bib. $14.95. Code: HARP

WEATHER WARFARE
The Military's Plan to Draft Mother Nature
by Jerry E. Smith

Weather modification in the form of cloud seeding to increase snow packs in the Sierras or suppress hail over Kansas is now an everyday affair. Underground nuclear tests in Nevada have set off earthquakes. A Russian company has been offering to sell typhoons (hurricanes) on demand since the 1990s. Scientists have been searching for ways to move hurricanes for over fifty years. In the same amount of time we went from the Wright Brothers to Neil Armstrong. Hundreds of environmental and weather modifying technologies have been patented in the United States alone – and hundreds more are being developed in civilian, academic, military and quasi-military laboratories around the world *at this moment!*

304 Pages. 6x9 Paperback. Illustrated. Bib. $18.95. Code: WWAR

MIND CONTROL, WORLD CONTROL
The Encyclopedia of Mind Control
by Jim Keith

Keith uncovers a surprising amount of information on the technology, experimentation and implementation of Mind Control technology. Various chapters in this shocking book are on early C.I.A. experiments such as Project Artichoke and Project RIC-EDOM, the methodology and technology of implants, Mind Control Assassins and Couriers, various famous "Mind Control" victims such as Sirhan Sirhan and Candy Jones. Also featured in this book are chapters on how Mind Control technology may be linked to some UFO activity and "UFO abductions.

256 Pages. 6x9 Paperback. Illustrated. References. $14.95. Code: MCWC

MIND CONTROL AND UFOS
Casebook on Alternative 3
by Jim Keith

A revised and updated edition of *Casebook on Alternative 3,* Keith's classic investigation of the Alternative 3 scenario as it first appeared on British television over 20 years ago. Keith delves into the bizarre story of Alternative 3, including mind control programs, underground bases not only on the Earth but also on the Moon and Mars, the real origin of the UFO problem, the mysterious deaths of Marconi Electronics employees in Britain during the 1980s, the Russian-American superpower arms race of the 50s, 60s and 70s as a massive hoax, more.

248 Pages. 6x9 Paperback. Illustrated. $14.95. Code: MCUF

MASS CONTROL
Engineering Human Consciousness
by Jim Keith

Conspiracy expert Keith's final book on mind control, Project Monarch, and mass manipulation presents chilling evidence that we are indeed spinning a Matrix. Keith describes the New Man, where conception of reality is a dance of electronic images fired into his forebrain, a gossamer construction of his masters, designed so that he will not perceive the actual. His happiness is delivered to him through a tube or an electronic connection. His God lurks behind an electronic curtain; when the curtain is pulled away we find the CIA sorcerer, the media manipulator… Chapters on the CIA, Tavistock, Jolly West and the Violence Center, Guerrilla Mindwar, Brice Taylor, other recent "victims," more.

256 Pages. 6x9 Paperback. Illustrated. Index. $16.95. code: MASC

PIRATES & THE LOST TEMPLAR FLEET
The Secret Naval War Between the Templars & the Vatican
by David Hatcher Childress

Childress takes us into the fascinating world of maverick sea captains who were Knights Templar (and later Scottish Rite Free Masons) who battled the ships that sailed for the Pope. The lost Templar fleet was originally based at La Rochelle in southern France, but fled to the deep fiords of Scotland upon the dissolution of the Order by King Phillip. This banned fleet of ships was later commanded by the St. Clair family of Rosslyn Chapel (birthplace of Free Masonry). St. Clair and his Templars made a voyage to Canada in the year 1298 AD, nearly 100 years before Columbus! Later, this fleet of ships and new ones to come, flew the Skull and Crossbones, the symbol of the Knights Templar.

320 PAGES. 6x9 PAPERBACK. ILLUSTRATED. BIBLIOGRAPHY. $16.95. CODE: PLTF

ORDER FORM

10% Discount When You Order 3 or More Items!

One Adventure Place
P.O. Box 74
Kempton, Illinois 60946
United States of America
Tel.: 815-253-6390 • Fax: 815-253-6300
Email: auphq@frontiernet.net
http://www.adventuresunlimitedpress.com

ORDERING INSTRUCTIONS

- ✓ Remit by USD$ Check, Money Order or Credit Card
- ✓ Visa, Master Card, Discover & AmEx Accepted
- ✓ Paypal Payments Can Be Made To: info@wexclub.com
- ✓ Prices May Change Without Notice
- ✓ 10% Discount for 3 or More Items

SHIPPING CHARGES

United States

- ✓ Postal Book Rate { $4.50 First Item; 50¢ Each Additional Item
- ✓ POSTAL BOOK RATE Cannot Be Tracked! Not responsible for non-delivery.
- ✓ Priority Mail { $6.00 First Item; $2.00 Each Additional Item
- ✓ UPS { $7.00 First Item; $1.50 Each Additional Item

NOTE: UPS Delivery Available to Mainland USA Only

Canada

- ✓ Postal Air Mail { $15.00 First Item; $2.50 Each Additional Item
- ✓ Personal Checks or Bank Drafts MUST BE US$ and Drawn on a US Bank
- ✓ Canadian Postal Money Orders OK
- ✓ Payment MUST BE US$

All Other Countries

- ✓ Sorry, No Surface Delivery!
- ✓ Postal Air Mail { $19.00 First Item; $6.00 Each Additional Item
- ✓ Checks and Money Orders MUST BE US$ and Drawn on a US Bank or branch.
- ✓ Paypal Payments Can Be Made in US$ To: info@wexclub.com

SPECIAL NOTES

- ✓ RETAILERS: Standard Discounts Available
- ✓ BACKORDERS: We Backorder all Out-of-Stock Items Unless Otherwise Requested
- ✓ PRO FORMA INVOICES: Available on Request
- ✓ DVD Return Policy: Replace defective DVDs only

ORDER ONLINE AT: www.adventuresunlimitedpress.com

10% Discount When You Order 3 or More Items!

Please check: ☑

☐ This is my first order ☐ I have ordered before

Name
Address
City
State/Province Postal Code
Country
Phone: Day Evening
Fax Email

Item Code	Item Description	Qty	Total

Please check: ☑

Subtotal ▶
Less Discount-10% for 3 or more items ▶

☐ Postal-Surface — Balance ▶
☐ Postal-Air Mail (Priority in USA) — Illinois Residents 6.25% Sales Tax ▶
Previous Credit ▶
☐ UPS (Mainland USA only) — Shipping ▶
Total (check/MO in USD$ only) ▶

☐ Visa/MasterCard/Discover/American Express

Card Number:
Expiration Date: Security Code:

✓ SEND A CATALOG TO A FRIEND: